Fine Resolution Remote Sensing of Species in Terrestrial and Coastal Ecosystems

Detailed and accurate information on the spatial distribution of individual species over large spatial extents and over multiple time periods is critical for rapid response and effective management of environmental change. The twenty-first century has witnessed a rapid development in both fine resolution sensors and statistical theories and techniques. These innovations hold great potential for improved accuracy of species mapping using remote sensing.

Fine Resolution Remote Sensing of Species in Terrestrial and Coastal Ecosystems is a collection of eight cutting-edge studies of fine spatial resolution remote sensing, including species mapping of biogenic and coral reefs, seagrasses, salt and freshwater marshes, and grasslands. The studies illustrate the power of fine resolution imagery for species identification, as well as the value of unmanned aerial vehicle (UAV) imagery as an ideal source of high-quality reference data at the species level. The studies also highlight the benefit of LiDAR (Light Detection and Ranging) data for species identification, and how this varies depending on the species of interest as well as the nature of the context in which the species is found. The broad range of applications explored in the book demonstrates the major contribution of remote sensing to species-level terrestrial and coastal ecosystem studies as well as the potential for future advances.

The chapters in this book were originally published as a special issue of the *International Journal of Remote Sensing*.

Qi Chen is Professor of Geography and Environment at the University of Hawai'i at Mānoa, Honolulu, USA. His research focuses on the use of LiDAR, high spatial resolution remote sensing, statistical modelling, and artificial intelligence for environmental mapping and monitoring.

Tiit Kutser is Professor of Remote Sensing at the Estonian Marine Institute at the University of Tartu, Tallinn, Estonia. His research covers many different topics from mapping water quality parameters (including harmful algal blooms) in coastal and inland waters to benthic habitat (including coral reefs) mapping and the role of lakes in the global carbon cycle.

Antoine Collin is Associate Professor of Geography and Ecology at the Paris Sciences & Letters (PSL) Research University, Dinard, France. His research links the coastal natural and social sciences in the ocean-climate change. He maps and models coastal environments using high spatio-temporal resolution spaceborne, airborne, handborne, waterborne data, and machine learning.

Timothy A. Warner is Emeritus Professor of Geology and Geography at West Virginia University, Morgantown, USA. He served as editor in chief of the *International Journal of Remote Sensing* from 2014 to 2020. He is a Fellow of the American Society of Photogrammetry and Remote Sensing.

Fine Resolution Remote Sensing of Species in Terrestrial and Coastal Ecosystems

Edited by
Qi Chen, Tiit Kutser, Antoine Collin and Timothy A. Warner

Routledge
Taylor & Francis Group

LONDON AND NEW YORK

First published 2022
by Routledge
2 Park Square, Milton Park, Abingdon, Oxon, OX14 4RN

and by Routledge
605 Third Avenue, New York, NY 10158

Routledge is an imprint of the Taylor & Francis Group, an informa business

Introduction, Chapters 1–5, 7 and 8 © 2022 Taylor & Francis

Chapter 6 © 2018 Dimitris Stratoulias, Heiko Balzter, András Zlinszky and Viktor R. Tóth. Originally published as Open Access.

British Library Cataloguing-in-Publication Data
A catalogue record for this book is available from the British Library

ISBN13: 978-1-032-04265-7 (hbk)
ISBN13: 978-1-032-04267-1 (pbk)
ISBN13: 978-1-003-19119-3 (ebk)

DOI: 10.4324/9781003191193

Typeset in Myriad Pro
by codeMantra

Publisher's Note
The publisher accepts responsibility for any inconsistencies that may have arisen during the conversion of this book from journal articles to book chapters, namely the inclusion of journal terminology.

Disclaimer
Every effort has been made to contact copyright holders for their permission to reprint material in this book. The publishers would be grateful to hear from any copyright holder who is not here acknowledged and will undertake to rectify any errors or omissions in future editions of this book.

Contents

Citation Information

The chapters in this book were originally published in the *International Journal of Remote Sensing*, volume 39, issue 17 (2018). When citing this material, please use the original page numbering for each article, as follows:

For any permission-related enquiries please visit:
http://www.tandfonline.com/page/help/permissions

Contributors

N. Balkenhol Centre of Biodiversity and Sustainable Land Use (CBL), University of Goettingen, Germany. Wildlife Sciences, Faculty of Forest Sciences and Forest Ecology, University of Goettingen, Germany.

Heiko Balzter Centre for Landscape and Climate Research, Department of Geography, University of Leicester, Leicester Institute for Space and Earth Observation, UK. National Centre for Earth Observation, University of Leicester, UK.

Elisa Casella ZMT, Leibniz Centre for Tropical Marine Research, Bremen, Germany.

Qi Chen Department of Geography and Environment, University of Hawai'i at Mānoa, Honolulu, USA.

Antoine Collin Ecole Pratique des Hautes Etudes, PSL Research University, CNRS LETG, Dinard, France.

Neil Davies Gump South Pacific Research Station, University of California, Moorea, French Polynesia. Berkeley Institute for Data Science, University of California, Berkeley, USA.

Sara Denka Department of Geosciences, Florida Atlantic University, Boca Raton, USA.

Stanislas Dubois IFREMER, Laboratoire d'Ecologie Benthique Côtière (LEBCO), Plouzané, France.

Benoît Espiau PSL Research University: EPHE-UPVD-CNRS, USR 3278 CRIOBE, Papetoai, Moorea, French Polynesia.

Samuel Etienne Ecole Pratique des Hautes Etudes, PSL Research University, Centre National de la Recherche Scientifique (CNRS – UMR LETG), Dinard, Brittany, France.

James L. Hench Nicholas School of the Environment, Duke University, USA.

Sally J. Holbrook Department of Ecology, Evolution and Marine Biology; Marine Science Institute; University of California Santa Barbara, USA.

J. Isselstein Grassland Science, Faculty of Agricultural Science, University of Goettingen, Germany. Centre of Biodiversity and Sustainable Land Use (CBL), University of Goettingen, Germany.

Ilmar Kotta Estonian Marine Institute, University of Tartu, Tallinn, Estonia.

Jonne Kotta Estonian Marine Institute, University of Tartu, Tallinn, Estonia.

Tiit Kutser Estonian Marine Institute, University of Tartu, Tallinn, Estonia.

Natasha Lambert Ecole Pratique des Hautes Etudes, PSL Research University, Centre National de la Recherche Scientifique (CNRS – UMR LETG), Dinard, Brittany, France.

Wahyu Lazuardi Cartography and Remote Sensing, Department of Geographic Information Science, Faculty of Geography, Universitas Gadjah Mada, Yogyakarta, Indonesia.

Franck Lerouvreur PSL Research University: EPHE-UPVD-CNRS, USR 3278 CRIOBE, Papetoai, Moorea, French Polynesia.

M. Meißner Institut für Wildbiologie Göttingen und Dresden e.V., Goettingen, Germany.

Deepak R. Mishra Center for Geospatial Research, Department of Geography, University of Georgia, Athens, USA.

Nao Nakamura LABoratoire d'EXcellence CORAIL, Perpignan, France. PSL Research University: EPHE-UPVD-CNRS, USR 3278 CRIOBE, Papetoai, Moorea, French Polynesia.

Helen Orav-Kotta Estonian Marine Institute, University of Tartu, Tallinn, Estonia.

Merli Pärnoja Estonian Marine Institute, University of Tartu, Tallinn, Estonia.

Yves Pastol Service Hydrographique et Océanographique de la Marine, Brest, Brittany, France.

Christoph Raab Grassland Science, Faculty of Agricultural Science, University of Goettingen, Germany. Centre of Biodiversity and Sustainable Land Use (CBL), University of Goettingen, Germany.

Camille Ramambason Ecole Pratique des Hautes Etudes (EPHE), PSL Research University, Dinard, Brittany, France.

N. Rohwer Institut für Wildbiologie Göttingen und Dresden e.V., Goettingen, Germany.

Alessio Rovere ZMT, Leibniz Centre for Tropical Marine Research, Bremen, Germany. Marum, University of Bremen, Germany.

Russell J. Schmitt Department of Ecology, Evolution and Marine Biology, and Marine Science Institute, University of California Santa Barbara, USA.

Gilles Siu PSL Research University: EPHE-UPVD-CNRS, USR 3278 CRIOBE, Papetoai, Moorea, French Polynesia.

Dimitris Stratoulias Balaton Limnological Institute, Centre for Ecological Research, Hungarian Academy of Sciences, Tihany, Hungary. Centre for Landscape and Climate Research, Department of Geography, University of Leicester, Leicester Institute for Space and Earth Observation, UK. GeoAnalysis, Budapest, Hungary. Singapore-MIT Alliance for Research and Technology (SMART), Singapore.

H. G. Stroh Grassland Science, Faculty of Agricultural Science, University of Goettingen, Germany.

Lauric Thiault LABoratoire d'EXcellence CORAIL, Perpignan, France. PSL Research University: EPHE-UPVD-CNRS, USR 3278 CRIOBE, Papetoai, Moorea, French Polynesia.

B. Tonn Grassland Science, Faculty of Agricultural Science, University of Goettingen, Germany.

Viktor R. Tóth Balaton Limnological Institute, Centre for Ecological Research, Hungarian Academy of Sciences, Tihany, Hungary.

Matthias Troyer Department of Physics, Institute for Theoretical Physics, ETH Zurich, Switzerland. Microsoft Quantum Research, Redmond, USA.

Ele Vahtmäe Estonian Marine Institute, University of Tartu, Tallinn, Estonia.

Timothy A. Warner Department of Geology and Geography, West Virginia University, Morgantown, USA.

Pramaditya Wicaksono Remote Sensing Laboratory, Department of Geographic Information Science, Faculty of Geography, Universitas Gadjah Mada, Yogyakarta, Indonesia.

Caiyun Zhang Department of Geosciences, Florida Atlantic University, Boca Raton, USA.

András Zlinszky Balaton Limnological Institute, Centre for Ecological Research, Hungarian Academy of Sciences, Tihany, Hungary.

INTRODUCTION

Fine resolution remote sensing of species in terrestrial and coastal ecosystems

Qi Chen, Tiit Kutser, Antoine Collin and Timothy A. Warner

Detailed and accurate information on the spatial distribution of individual species over large spatial extents and over multiple time periods is critical for rapid response and effective management of environmental change. Although remote sensing has been used to map species for decades, the long-standing challenge is that the accuracy of species maps is often too low to meet the requirements of management, or the methods are too complex or location-specific to be used in routine mapping. On the other hand, in the 21st century, we have witnessed a rapid development in both fine resolution remote sensors and statistical theories and techniques, which hold great potential for improved accuracy of species mapping.

This special issue is a collection of eight studies that present cutting-edge research on using fine resolution remotely sensed data for mapping species or communities in terrestrial and coastal ecosystems. Several important findings emerge from these studies:

(1) **Fine spatial resolution is particularly useful for species identification**. Conventional remote sensing has focused on the use of spectral characteristics for species identification. However, many studies in this special issue demonstrate the power of fine spatial information. Zhang, Denka, and Mishra (2018) found that the texture features from very high spatial resolution (30 cm (1 foot) resolution) aerial photography enhanced the overall accuracy of classifying freshwater marsh species by 14.2% compared with the use of spectral data alone. Collin, Lambert, and Etienne (2018b) found that pan-sharpened Wordview-3 (WV-3) imagery at 0.31 m resolution resulted in an overall accuracy of 95.47% for classifying 15 salt marsh habitats, in comparison to just 82.33% when using 1.24 m resolution imagery.

(2) **Unmanned aerial vehicle (UAV) imagery can provide high quality reference data at the species level**. Over heterogeneous areas, collecting reference data at the level of individual plants and individual species may require labor-intensive field mapping (Raab et al. 2018). Moreover, mixed pixels are common at fine spatial resolutions; even at the sub-meter level (e.g., 0.5 m by 0.5 m quadrats), the space is

not necessarily fully occupied with a single species (Collin et al. 2018a). UAV (or drone) imagery, due to its extremely high (centimeter-level) possible spatial resolution, is a cost-effective data source for providing reference information (which Collin et al. 2018a, 2018b, 2018c term 'air truth') over relatively large areas.

(3) **The value of LiDAR (Light Detection and Ranging) for species identification is specific to individual studies.** Several studies in this issue incorporated airborne LiDAR data, but the evaluation of the value of LiDAR for species classification was mixed. Collin et al. (2018c), using only the DSM (Digital Surface Model) and DIM (Digital Intensity Model) data, were able to map the coral reef species with an overall accuracy of 75%. Zhang, Denka, and Mishra (2018) found that the addition of DTM (Digital Terrain Model) features improved the average overall accuracy of species classification by 4.3%, compared to the use of aerial imagery alone. However, Stratoulias et al. (2018) found that the incorporation of a LiDAR-derived Digital Height Model (DHM) to airborne hyperspectral imagery did not improve the accuracy of classifying emergent wetland vegetation types in Hungary. This indicates that the value of LiDAR for species identification is specific to individual studies that extract different variables and use different types of data over different locations.

(4) **Species-level mapping is one of the major challenges in remote sensing and research on the topic is still in its infancy.** Accurate and effective mapping of species ideally requires high spectral resolution, high spatial resolution, high temporal resolution, and the use of multiple data sources. Conventionally, trade-offs exist between these requirements. Recent advances in remote sensors and platforms, including imaging hyperspectral spectroradiometers (Vahtmäe et al. 2018) and low-cost high spatial and temporal resolution PlanetScope imagery (Wicaksono and Lazuardi 2018), are at least partially overcoming these obstacles. The studies compiled in this special issue are just the beginning of an era of making detailed species-level mapping over large areas for better management of a rapidly changing planet.

References

Collin, A., S. Dubois, C. Ramambason, and S. Etienne. 2018a. "Very High Resolution Mapping of Emerging Biogenic Reefs Using Airborne Optical Imagery and Neural Network: The Honeycomb Worm (Sabellaria Alveolata) Case Study." *International Journal of Remote Sensing* 39: 5660–5675. doi:10.1080/01431161.2018.1484964.

Collin, A., N. Lambert, and S. Etienne. 2018b. "Satellite Based Salt Marsh Elevation, Vegetation Height, and Species Composition Mapping Using the Superspectral WorldView-3 Imagery." *International Journal of Remote Sensing* 39: 5619–5637. doi:10.1080/01431161.2018.1466084.

Collin, A., C. Ramambason, Y. Pastol, E. Casella, A. Rovere, L. Thiault, B. Espiau, et al. 2018c. "Very High Resolution Mapping of Coral Reef State Using Airborne Bathymetric LiDAR Surface-Intensity and Drone Imagery." *International Journal of Remote Sensing*: 1–13 39: 5676–5688. doi:10.1080/01431161.2018.1500072.

Raab, C., H. Stroh, B. Tonn, M. Meißner, N. Rohwer, N. Balkenhol, and J. Isselstein. 2018. "Mapping Semi-Natural Grassland Communities Using Multitemporal RapidEye Remote Sensing Data." *International Journal of Remote Sensing* 39: 5638–5659. doi: 10.1080/01431161.2018.1504344.

Stratoulias, D., H. Balzter, A. Zlinszky, and V. R. Tóth. 2018. "A Comparison of Airborne Hyperspectral-Based Classifications of Emergent Wetland Vegetation at Lake Balaton, Hungary." *International Journal of Remote Sensing* 39: 5689–5715. doi:10.1080/01431161.2018.1466081.

Vahtmäe, E., J. Kotta, H. Orav-Kotta, I. Kotta, M. Pärnoja, and T. Kutser. 2018. "Predicting Macroalgal Pigments (Chlorophyll A, Chlorophyll B, Chlorophyll a + B, Carotenoids) in Various Environmental Conditions Using High-Resolution Hyperspectral Spectroradiometers." *International Journal of Remote Sensing* 39: 5716–5738. doi:10.1080/01431161.2017.1399481.

Wicaksono, P., and W. Lazuardi. 2018. "Assessment of PlanetScope Images for Benthic Habitat and Seagrass Species Mapping in a Complex Optically-Shallow Water Environment." *International Journal of Remote Sensing* 39: 5739–5765. doi:10.1080/01431161.2018.1506951

Zhang, C., S. Denka, and D. R. Mishra. 2018. "Mapping Freshwater Marsh Species in the Wetlands of Lake Okeechobee Using Very High-Resolution Aerial Photography and Lidar Data." *International Journal of Remote Sensing* 39: 5600–5618. doi:10.1080/01431161.2018.1455242.

Mapping freshwater marsh species in the wetlands of Lake Okeechobee using very high-resolution aerial photography and lidar data

Caiyun Zhang, Sara Denka and Deepak R. Mishra

ABSTRACT

Accurate marsh species maps are needed to study their abundance, distribution, habitat change, and to support the ongoing and future restoration activities in Lake Okeechobee, Florida, U.S.A. In this study, we integrated very high-resolution aerial photography with a light detection and ranging (LiDAR)-derived digital elevation model (DEM) for the discrimination of six freshwater marsh species (*Salix caroliniana, Spartina bakeri, Polygonum* spp., *Typha* spp., *Phragmites australis*, and *Cladium jamaicense*). Four techniques were combined in the mapping procedure, including object-based image analysis, machine learning classifiers, texture analysis, and ensemble analysis. The results showed that both texture and topography features were invaluable for marsh species mapping. A synergy of spatial, spectral, and topographical features achieved an overall accuracy (OA) of 85.3% and kappa coefficient of 0.83 using the Random Forest (RF) classifier. Ensemble analysis of the outputs from Support Vector Machine (SVM), RF, and Artificial Neural Network (ANN) did not increase the classification accuracy, but produced an uncertainty map to identify regions with a robust classification and areas difficult to map. Ensemble analysis was beneficial in getting further insight into marsh species mapping. The developed digital procedure is a promising alternative to traditional field survey and manual interpretation methods for generating marsh maps to support the lake restoration.

1. Introduction

1.1. *Significance of marsh mapping in Lake Okeechobee*

Lake Okeechobee, Florida, is the largest freshwater lake in the Southeastern United States. It covers 730 square miles (1890.69 km^2) and is known as *Florida's Inland Sea*. The lake and its wetlands are at the centre of a much larger watershed, the Greater Everglades, which stretches from the Kissimmee River to Florida Bay (Figure 1). Lake Okeechobee is a key component of South Florida's water supply and flood control system, and it functions as the heart of the Greater Everglades. The original lake

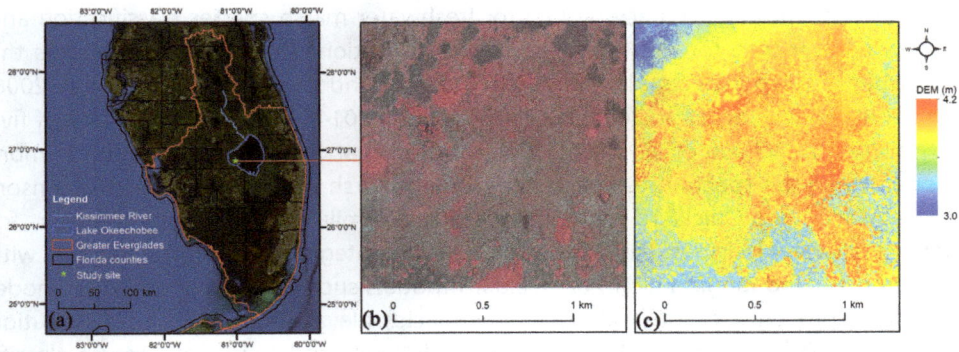

Figure 1. Location of the Greater Everglades, and Lake Okeechobee (a), and the study site shown as a colour infrared (CIR) composite of the 1-foot (30.48 cm) resolution aerial photography collected on 24 April 2012 (b), and the 1-foot (30.48 cm) DEM generated from lidar data collected in 2007 (c).

ecosystem was severely modified over the past century by human activities, resulting in many environmental issues in South Florida. A large amount of phosphorus accumulated in lake sediments due to excessive phosphorus loads from the lake's watershed. Large lake freshwater discharges harmed the ecological health of downstream estuaries (Lake Okeechobee Watershed Restoration Project 2018). The Lake Okeechobee Watershed Restoration Project is currently underway to improve the water quality and quantity, timing of flows, and restoration of the lake's wetlands. This project is part of the Comprehensive Everglades Restoration Plan (CERP) which is the largest hydrological restoration effort in the U.S.A. with an estimated cost of more than $10.5 billion and a 35 + year timeline (CERP 2018). CERP requires vegetation maps at different scales because these restoration efforts will cause modification to plant communities and species (Doren, Rutchey, and Welch 1999). For the lake area, detailed marsh maps are needed to assess the effects and progress of the restoration.

1.2. *Past efforts in marsh mapping*

Currently, marsh maps in Lake Okeechobee are mainly generated through manual interpretation of large-scale aerial photography using stereo techniques. In the stereo analysis, a grid is created and superimposed on the georeferenced three-dimensional stereo imagery first, and then a photo-interpreter manually labels each grid using a Softcopy Photogrammetry Workstation. Only a certified photogrammetrist is allowed to carry out the manual interpretation which is time consuming and labour intensive. Inconsistent results can be produced by different mapping personnel. With the advance of digital image processing techniques, it is expected that this manual procedure can be semi-automated or automated. Numerous efforts have been made to map fresh and salt marshes in the past decades, which can be broadly grouped into three categories based on the application of data sources. The first category is the application of multispectral imagery such as data from WorldView, QuickBird, and aerial photography (e.g. Belluco et al. 2006; Carle, Wang, and Sasser 2014; Kumar and Sinha 2014; Sun et al. 2016). Results have shown that fine spatial resolution multispectral imagery is useful for mapping marsh species, with an accuracy of 75% for freshwater marsh species classification and accuracy of over 90% for salt marsh species classification. The second category is the employment of hyperspectral imagery (e.g. Li, Ustin, and Lay 2005; Belluco et al. 2006; Wang et al. 2007; Judd et al. 2007; Kumar and Sinha 2014). A comparison study of five data sources by Belluco et al. (2006) demonstrated that the spatial resolution is more important than the spectral resolution in mapping marsh species; multispectral sensors such as IKONOS and QuickBird can achieve accuracy similar to hyperspectral sensors for mapping saltwater marshes. The third category is the integration of optical imagery with light detection and ranging (lidar) data. Lidar products such as a digital elevation model (DEM) have proven useful to extract marsh species elevation ranges and distribution (Morris et al. 2005; Sadro, Gastil-Buhl, and Melack 2007), and to improve marsh classification (Gilmore et al. 2008; Hladik, Schalles, and Alber 2013; Mishra et al. 2015). Lidar is able to characterize plant structure and topography by collecting elevation data to complement the spectral information of optical imagery. Marsh species distribution is related to the hydroperiod of wetlands, which is correlated with the ground elevation. This indicates that lidar has the potential to increase marsh species classification. Application of multi-sensor and multi-temporal data through data fusion techniques is promising for marsh species mapping and characterization. However, studies in this context are limited.

1.3. *Objectives of this study*

The main objective of this study is to evaluate whether a machine learning framework by integrating very high-resolution aerial photography (30.48 cm) with a lidar-DEM is suitable for marsh species mapping in the wetlands of Lake Okeechobee. Very high-resolution aerial photography (1 m or smaller) has been frequently collected over the lake area to support ongoing restoration. But examination of this type of data using digital techniques is limited. An integration of multiple data sources for this purpose is even scarcer.

Past efforts in marsh species mapping focus on the application of pixel-based analysis and traditional classifiers such as Maximum Likelihood. In this study, we combined contemporary digital processing techniques for marsh species mapping, including object-based image analysis (OBIA), machine learning classifiers, ensemble analysis, and texture analysis. OBIA is more useful than pixel-based analysis for classifying fine spatial resolution imagery and it has been applied in various fields (Blaschke 2010). Object-based mapping is desirable for heterogeneous regions in the Greater Everglades because pixel-based methods may lead to the 'salt-and-pepper' effect (Zhang and Xie 2012, 2013, 2014; Zhang 2014; Zhang, Xie, and Selch 2013; Zhang, Selch, and Cooper 2016). Machine learning classifiers such as Support Vector Machine (SVM), Random Forest (RF), and Artificial Neural Network (ANN) have proven powerful for mapping vegetation communities in the Greater Everglades. When data fusion is applied in the classification, the distribution of the data for each class commonly does not follow a Gaussian distribution, which may lead to a poor performance of the parametric classifiers, such as Maximum Likelihood. Non-parametric machine learning classifiers are thus preferred when a complex dataset is used in the classification. In addition, these classifiers have not been comprehensively assessed and compared for marsh species mapping. Recent studies have demonstrated that an ensemble analysis of multiple classifications can improve or generate a more robust result for mapping complex wetlands (Zhang 2014; Zhang, Selch, and Cooper 2016), benthic habitats in coastal environments (Zhang 2015), and urban land cover types (Zhang, Smith, and Fang 2018). Ensemble analysis is a multiple classification system that integrates the outputs of several classifiers to improve the classification or make the classification more robust compared with the application of one classifier. It is expected that ensemble analysis is also valuable for marsh species mapping, but this prospect has not been evaluated before. Texture features extracted from fine resolution imagery have also proven valuable for mapping plant communities in the Greater Everglades (Zhang and Xie 2012; Szantoi et al. 2013; Szantoi et al. 2015). But the contribution of texture features for marsh species classification has not been evaluated. Research combining multiple data sources and multiple digital processing techniques (OBIA, machine learning, ensemble analysis, and texture analysis) is rare in marsh species mapping. To this end, the specific objectives of this study are: (1) to evaluate whether integration of aerial photography and a lidar-DEM can improve the accuracy of marsh species classification, compared with using fine resolution aerial photography alone, (2) to assess the contribution of texture features in marsh species mapping, and (3) to examine the capability of machine learning classifiers and the benefit of ensemble analysis for characterizing the marsh ecosystem.

2. Study site and data

2.1. *Study site*

The study area is located at the western edge of Lake Okeechobee, a region known as the Moore Haven Marsh (Figure 1). The spatial distribution of marsh species over this region is mainly determined by the hydroperiod (i.e. frequency of inundation, as percentage of days per year) with short hydroperiod regions supporting species such as spikerush and willow, and long hydroperiod regions supporting species such as cattail and sawgrass (Havens and Gawlik 2005). Freshwater graminoid marsh and floating emergent marsh are two dominant marsh communities in the wetlands of Lake Okeechobee. The selected site is a 574-acre (2.32 km^2) portion of the wetlands. It is dominated by six freshwater species to be identified in this study, including Willow (*Salix caroliniana*), Cordgrass (*Spartina bakeri*), Smartweed (*Polygonum* spp.), Cattail (*Typha* spp.), Common reed (*Phragmites australis*), and Sawgrass (*Cladium jamaicense*). These species are common species of graminoid marsh over this region. In addition, Meadow Marsh (Graminoid), Swamp Shrubland, and Mud are also present in the study site and were mapped. The Graminoid marsh is characterized by emergent herbaceous vegetation and is a mosaic of various grasses. If no dominant marsh species can be identified, the region is then labelled as Graminoid marsh. Swamp shrubland is seasonally to semi-permanently flooded, high-density stands of small trees or shrubs with heights less than 5 m and is found throughout Florida. Mud refers to moist, open ground.

2.2. *Data*

Data used in this study included 1-foot (30.48 cm) resolution aerial photography, reference data, and airborne lidar data. The aerial photography was collected on 24 April 2012 using a Microsoft Vexcel UltraCamX Senor System as part of a project led by the South Florida Water Management District (SFWMD). The objective of this project was to produce high-resolution digital aerial imagery for the wetlands at Lake Okeechobee Moore Haven Marsh. The acquired raw images were geometrically and radiometrically processed, and orthorectified into 1-foot (30.48 cm) orthos with four bands (Red, Green, Blue, and near-infrared). A total of 223 tiles (5000 ft/ 1524 m × 5000 ft/1524 m per tile) were generated for the project. Only one tile was selected based on the reference data to test the digital procedure developed in this study (Figure 1(b)). The selected tile covers the dominant species present in the marsh area. Stereo imagery was captured with 60% overlap between adjacent frames and 40% sidelap between adjacent flight lines. The stereo imagery was manually interpreted using a stereo plotter to generate a marsh map by the SFWMD. The features were labelled based on a classification system developed by Rutchey et al. (2006) for vegetation classification of south Florida natural areas. The classification system is hierarchical, designating up to six levels with lower levels being nested within higher levels. Graminoid and/or herbaceous emergent or floating vegetation in shallow water that stands at or above the ground surface for much of the year are identified as marsh. Marsh is further divided into salt marsh and freshwater marsh (level 2). For the freshwater marsh, it is further detailed to graminoid marsh and floating emergent marsh (level 3). Species labelling belongs to level 4. During manual interpretation, each grid cell is labelled with the majority vegetation category observed. Level 4 is further detailed by adding more modifications such as density, leading to levels 5 and 6. All levels are labelled in the manual procedure. The manually interpreted map was provided by the SFWMD and used as a reference dataset for this study. The map was reported to have a minimum accuracy of 90% based on extensive field surveys.

Airborne lidar data was collected between June and December 2007 by Merrick & Company (Greenwood Village, CO, U.S.A.) using a Leica ALS-50 system to support the Florida Division of Emergency Management. There is a 5-year time gap between the acquisition of the lidar data (2007) and aerial photography (2012). The Leica ALS-50 lidar system collects small footprint multiple returns and intensity at 1060 nm wavelength. The vendor reported that the positional accuracy was 0.05 ft (1.524 cm) horizontally and 0.2 ft (6.096 cm) vertically at a 95% confidence level. The averaged point density for the study site is 2.8 points/m^2. The lidar point cloud data was processed and classified by the vendor using the Merrick Advanced Remote Sensing (MARS) processing software package.

3. Methodology

Three data fusion techniques are commonly used to combine optical imagery and lidar data for classification, including pixel-level fusion, feature-level fusion, and decision-level fusion (Gómez-Chova et al. 2015). We applied a feature/decision-level fusion method (Zhang 2014; Zhang, Smith, and Fang 2018) to combine two data sources (aerial photography and lidar), and four data processing techniques (OBIA, machine learning classifiers, texture analysis, and ensemble analysis) for species mapping. This feature/ decision-level fusion method is useful to combine hyperspectral and lidar data for wetland plant community mapping (Zhang 2014), and urban land cover type mapping (Zhang, Smith, and Fang 2018).

Multiple steps are required in the mapping procedure, as shown in Figure 2. We first produced a 1-foot (30.48 cm) lidar-DEM to represent the topography of the bare terrain from the lidar ground points using the Inverse Distance Weighted interpolation technique. The lidar-DEM is shown in Figure 1(c). Note that the digital surface model representing the elevation of non-ground features from lidar was not produced or used in this study because of the large time gap between the acquisition of the lidar data and aerial photography. The spatial distribution of marsh plants is dynamic within a year and between years, while the topographical change might be small. We thus only included the lidar-DEM in the mapping procedure. To conduct object-based mapping, image objects were first produced through image segmentation, then spatial (texture) and spectral features were extracted to combine with topographical features (statistical descriptors) from the lidar-DEM at the object level to generate a fused dataset. This is known as feature-level fusion in multi-sensor data fusion applications (Gómez-Chova et al. 2015). Three machine learning classifiers (SVM, RF, and ANN) were used to pre-classify the fused dataset. The final outcome was derived through ensemble analysis of the three classifications using a decision-level fusion strategy. Note that the decision-level fusion strategy was based on the classification results of the fused dataset, rather than making a decision from the classifications of the individual data sources. The latter is commonly referred to as decision-level fusion in the literature. Consequently, an object-based marsh map was generated and evaluated by standard accuracy assessment

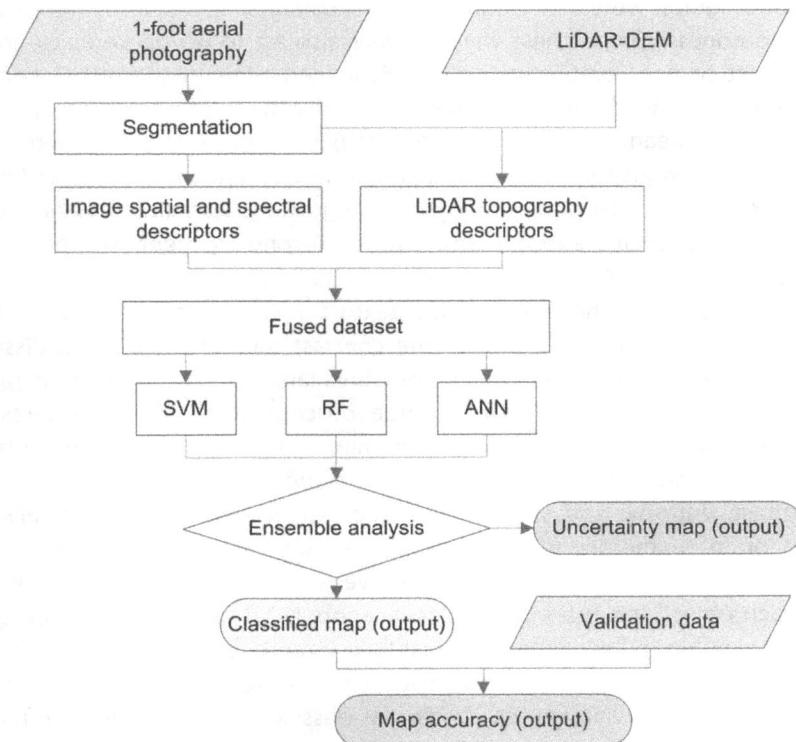

Figure 2. Methodology flow chart for mapping marsh species by combining 1-foot (30.48 cm) resolution aerial photography and lidar-DEM data.

approaches. Both feature-level fusion and decision-level fusion were used in the mapping procedure, and thus the fusion strategy was called the feature/decision-level fusion. The ensemble analysis of three classifications also produces an uncertainty map which defines the confidence of the classification for each image object. Major steps in the mapping procedure included image segmentation, classification, ensemble analysis, and accuracy assessment.

3.1. *Image segmentation*

The multi-resolution segmentation algorithm in eCognition Developer 9.0 (Trimble 2014) was used to generate image objects from the 1-foot (30.48 cm) aerial photography. This algorithm requires the inputs of several parameters, including scale, colour/shape, and smoothness/compactness, among which the scale parameter is the most important input. Efforts have been made to develop approaches for the optimal scale parameter determination (e.g. Grybas, Melendy, and Congalton 2017). Here we applied an unsupervised image segmentation evaluation approach developed by Johnson and Xie (2011) to determine an optimal scale parameter. This approach begins with a series of segmentations by setting different scale parameters, and then identifies the optimal image segmentation using a method that takes into account global intra-segment and inter-segment heterogeneity measures. Application of this approach to the aerial photograph of the study site revealed that a scale of 75 was optimal. All four spectral bands of the aerial photograph were set to equal weights, colour/shape weights were set to 0.5/0.5, and smoothness/compactness weights were also set to 0.5/0.5 so as to not favour either compact or non-compact segments. Following segmentation, the spectral and texture features of the aerial photography and lidar-DEM statistical descriptors (maximum, minimum, mean, and standard deviation) were extracted and merged at the object level as the input for classification. We extracted first- and second-order metrics for each band of the aerial photography including mean, standard deviation, contrast, dissimilarity, homogeneity, entropy, and angular second moment. The grey level co-occurrence matrix (GLCM) algorithm was used to extract the second-order texture measures by calculating the mean of the texture results in all four directions. Hall-Beyer (2017) suggests choosing one of the contrast measures (contrast, dissimilarity, and homogeneity), one of the orderliness measures (angular second moment, maximum probability, and entropy), and two or three descriptive statistics measures (mean, variance, and correlation) metrics, because many texture measures are intrinsically similar. We thus only applied first-order mean, standard deviation, and second-order GLCM contrast, entropy, and standard deviation in the classification. Details for the calculation of these metrics for each object can be found in the work by Trimble (2014). In addition, the normalized difference vegetation index (NDVI) was also calculated for each object and used in the classification. NDVI has proven useful for marsh community mapping in Everglades National Park (Szantoi et al. 2013).

After segmentation, a total of 536 image objects were selected as the reference objects to calibrate and validate the developed classification procedure. We followed a spatially stratified data sampling strategy, in which a fixed percentage of samples were selected for each class. The number of samples for each class was roughly estimated based on the results of image segmentation and the reference dataset. For example, a total of around 350 sawgrass objects are produced for the study site. If we select 20% of sawgrass objects as the reference objects, around 70 sawgrass objects will be randomly selected over the study site. However, if a species covers a small region with a limited number of objects produced, a small number of reference samples will be generated using the fixed percentage strategy, which might make the accuracy assessment statistically unreliable. For such a case, a higher percentage should be used in reference sample selection for this species.

3.2. *Classification: SVM, RF, and ANN*

Three machine learning classifiers, SVM, RF, and ANN, were used to pre-classify the fused dataset for discriminating six marsh species and three other land cover types. They have proven powerful for mapping plant communities in the Greater Everglades either to classify the fused dataset (e.g. Zhang, Xie, and Selch 2013; Zhang 2014) or to classify aerial photography alone (e.g. Szantoi et al. 2013; 2015). SVM is a supervised classifier with the aim to find a hyperplane that can separate the input dataset into a discrete predefined number of classes in a fashion consistent with the training samples (Vapnik 1995). It has been applied in various fields, as reviewed by Mountrakis, Im, and Ogole (2011). RF is a decision-tree-based ensemble classifier. Detailed descriptions of RF can be found in the work by Breiman (2001) and a recent review of RF in remote sensing was given by Belgiu and Drăguţ (2016). ANN is also an important technique in image classification. Various ANN algorithms have been developed and applied, as reviewed by Mas and Flores (2008). Here we applied the multilayer perceptron algorithm of ANN. All three classifiers require the inputs of several parameters. For example, SVM needs to set the kernel functions, and RF needs to set the number of selected variables for splitting each node in a tree. In this study, each classifier was implemented and tuned in Waikato Environment for Knowledge Analysis (WEKA), a machine learning open source software package (Hall et al. 2009). WEKA has an experimenter function which can effectively tune an algorithm by changing the parameters required by this algorithm and find the best model based on the statistical metrics. WEKA integrates more than 20 classifiers. Here we only selected three algorithms which produced comparable accuracies to be used in ensemble analysis.

3.3. *Ensemble analysis*

The innovation of this study is the use of ensemble analysis in the marsh mapping procedure. Ensemble analysis has proven valuable for plant community mapping in the Greater Everglades (Zhang 2014; Zhang and Xie 2014; Zhang, Selch, and Cooper 2016). This approach is a multiple classification system that combines the outputs of several classifiers. The classifiers in the system should generally produce accurate results but show some differences in per-class accuracy (Du et al. 2012). A strategy is required in ensemble analysis to integrate multiple classifications. The majority vote strategy is straightforward and commonly used in the literature. Each individual classifier with equal weights votes for an unknown input object without considering their performances on each individual class. A weighting strategy may improve the performance of the majority vote by weighting each classifier based on its accuracies for the reference data. A combination of the majority vote and the weighting strategy is more useful compared with the application of the majority vote alone (Zhang, Selch, and Cooper 2016). We thus applied this strategy in the ensemble analysis. If three votes are different for an unknown object, then the unknown object will be assigned to the class which has the highest accuracy among the classifiers. That is, the classifier with the best performance will obtain a weight of 1, while weights of other classifiers will be set at 0. If two or three classifiers vote the same class for an input object, then the object will be assigned to the same voted class. If three classifiers vote the same class for an unknown object, a full agreement will be achieved. Conversely if three votes are completely different, no agreement will be obtained. If two classifiers vote for the same class, a partial agreement will be produced. Consequently, an uncertainty map can be produced in conjunction with the final classified map from the ensemble analysis.

3.4. *Accuracy assessment*

In this study, the *k*-fold cross validation technique was used in the training and testing procedure. This evaluation method has proven valuable in machine learning classifications (Anguita et al. 2012). It splits the reference data into *k*-subsets first, and then iteratively, some of them are used to train the model and the others are exploited to assess model performance. The variable *k* was commonly set to 10 in the literature, which was also used for this study. After the iteration, classes are predicted for all input reference objects, which can then be used to calculate the error matrix and kappa statistics in terms of the true classes of the reference objects. McNemar test (Foody 2004) was also applied in the study to evaluate the statistical significance of differences in accuracy between different classifications. The difference in accuracy of a pair of classifications is viewed as being statistically significant at a confidence of 95% if the *Z*-score is larger than 1.96.

4. Results

4.1. *Experimental analysis using different datasets and classifiers*

To evaluate the contribution of texture features and a lidar-DEM for marsh mapping, a total of nine experiments were designed. Experiments 1–3 were applied to the aerial photograph excluding texture features (referred to as Dataset 1), i.e. only four spectral bands and NDVI were included in the classification; experiments 4–6 were applied to the aerial photograph including texture features (referred to as Dataset 2), i.e. spectral, NDVI, and texture features were included; and experiments 7–9 were applied to the fused dataset which combined the spectral, NDVI, texture, and topographical features (referred to as Dataset 3). The overall accuracies (OAs) and kappa coefficients from these experiments are displayed in Table 1. For Dataset 1, the SVM classifier obtained the highest accuracy with an OA of 67.2% and kappa coefficient of 0.62. For Dataset 2, when texture features were added, the OA was improved to 81.0% and kappa coefficient was increased to 0.78 by the ANN classifier which produced the highest accuracy among three classifiers. For Dataset 3 which included lidar-DEM features, the highest accuracy was achieved using the RF classifier with an OA of 85.3% and kappa coefficient of 0.83. Kappa tests showed that all the classifications from experiments 1–9 were statistically

Table 1. Classification accuracies and results of the statistical tests from different datasets and classifiers. The highest accuracy among three classifiers for each dataset is in bold.

Experiments	OA (%)	Kappa coefficient	Z-score (kappa coefficient)	Z-score (McNemar)
		Dataset 1: aerial photography excluding texture		
1. SVM	**67.2**	**0.62**	26.16	1.11 (1/2)
2. RF	65.1	0.59	24.77	0.40 (2/3)
3. ANN	65.9	0.60	25.30	0.98 (1/3)
		Dataset 2: aerial photography including texture		
4. SVM	80.0	0.77	38.50	0.00, 6.27[a] (4/5, 1/4)
5. RF	80.0	0.77	38.61	0.80, 6.86[a] (5/6, 2/5)
6. ANN	**81.0**	**0.78**	39.79	0.78, 7.30[a] (4/6, 3/6)
		Dataset 3: a combination of aerial photography and lidar-DEM		
7. SVM	84.9	0.83	46.03	0.32, 3.68[a] (7/8, 4/7)
8. RF	**85.3**	**0.83**	46.78	1.72, 4.43[a] (8/9, 5/8)
9. ANN	83.2	0.81	43.03	1.44, 1.81 (7/9, 6/9)
10. EA	85.1	0.83	46.43	0.23, 0.22, 1.54 (7/10, 8/10, 9/10)

SVM: Support Vector Machine; RF: Random Forest; ANN: Artificial Neural Network; OA: overall accuracy.
For the McNemar tests, 1/2, 1/3, 2/3, ..., 9/10 refer to the test between experiments 1 and 2, 1 and 3, 2 and 3, ..., 9 and 10, respectively; [a]Significant with 95% confidence.

better than a random classification. McNemar tests revealed that there was no significant difference between the classifiers if the same dataset was used; and the improvement from texture features was statistically significant (Table 1). The classifications from SVM and RF showed that the improvement from the lidar-DEM features was also statistically significant. ANN showed an increase in the OA and kappa coefficient using Dataset 3 compared with the application of Dataset 2, but the McNemar test demonstrated that the improvement from the lidar-DEM was not statistically significant. In general, the best result was from the fused dataset with the OAs ranging from 83.2% to 85.3% and kappa coefficients from 0.81 to 0.83 using three classifiers.

4.2. Ensemble analysis

The per-class accuracies from experiments 1–9 are displayed in Table 2 (here we only present the producer's accuracies). Discrimination of all eight marshes was improved by including texture features, in particular, for the Cattail classification. ANN and SVM completely failed to identify Cattail if texture features were excluded. When the lidar-DEM features were included, three classifiers showed a consistent increase in classifying five marshes: Cordgrass (*Spartina bakeri*), Smartweed (*Polygonum* spp.), Cattail (*Typha* spp.), Sawgrass (*Cladium jamaicense*), and Swamp Shrubland. In contrast, they produced uneven accuracy changes in identifying the other four classes (Table 2).

The experiments demonstrated that the fused dataset should be selected for the final marsh map because it achieved the highest accuracy among three datasets. Three classifiers produced a comparable result for the fused dataset; however, further examination showed that their performance was not completely even in identifying each individual class. For example, SVM produced the best result in characterizing classes 5–7; RF had the best performance in discriminating classes 4 and 8; and ANN showed the highest accuracy in classifying classes 2 and 3. This diversity is mainly caused by the discrepancies in three algorithms. SVM looks for a hyperplane to categorize data. RF explores the optimal decision tree to group data, whereas ANN trains the codebooks through iterative learning to classify data. The diversity drove us to explore the

Table 2. Per-class accuracies (%) from different datasets and classifiers. The highest accuracy among three datasets using each classifier is in bold.

Classes	SVM			RF			ANN		
	D1	D2	D3	D1	D2	D3	D1	D2	D3
1. Willow (*Salix caroliniana*)	94.4	98.1	98.1	93.0	96.2	**98.1**	87.7	**100.0**	96.3
2. Cordgrass (*Spartina bakeri*)	57.6	84.8	**87.0**	62.2	78.7	**87.0**	57.1	79.2	**87.5**
3. Smartweed (*Polygonum* spp.)	70.7	72.7	**77.4**	55.6	69.0	**75.9**	50.0	71.7	**78.7**
4. Cattail (*Typha* spp.)	0.0	66.7	**77.8**	28.6	46.7	**78.6**	0.0	50.0	**66.7**
5. Common reed (*Phragmites australis*)	88.5	98.6	98.6	93.3	95.9	**97.2**	89.7	**94.7**	94.6
6. Sawgrass (*Cladium jamaicense*)	52.9	64.9	**84.0**	44.8	70.0	**81.9**	54.8	74.0	**77.9**
7. Graminoid	52.3	74.6	**79.4**	46.6	**78.1**	77.2	53.4	75.6	**76.4**
8. Swamp shrubland	58.9	76.4	**77.8**	59.2	77.3	**80.8**	56.8	76.3	**76.4**
9. Mud	**100.0**	95.8	89.3	96.3	96.3	**100.0**	96.2	100.0	100.0

D1: Dataset 1; D2: Dataset 2; D3: Dataset 3, as shown in Table 1; SVM: Support Vector Machine; RF: Random Forest; ANN: Artificial Neural Network.

ensemble analysis of three outputs. The ensemble analysis result is shown in Table 1 as experiment 10 with an OA of 85.1% and kappa coefficient of 0.83. McNemar tests were also conducted between the classifications using ensemble analysis and classifications from experiments 7–9. The McNemar Z-scores showed that there was no significant difference in classification between ensemble analysis, SVM, RF, and ANN (Table 1).

4.3. *Object-based marsh mapping and uncertainty mapping*

Landis and Koch (1977) suggested that kappa coefficients larger than 0.80 indicate a strong agreement or accuracy. Classifications of the fused dataset from three classifiers and ensemble analysis produced a strong agreement with kappa coefficients of 0.81 (ANN) and 0.83 (SVM, RF, and ensemble analysis), respectively. The object-based marsh maps were thus produced using three classifiers and ensemble analysis, as shown in Figure 3. In general, the classified maps showed a consistent spatial distribution of marshes for the selected study site, which was dominated by Graminoid, Sawgrass, Cordgrass, and Swamp Shrubland. Cattail and Smartweed were sparse; Common Reed and Willow were moderate and shown as bigger patches (Figure 3(a–d)). An error matrix was also constructed for the classified map from ensemble analysis, as shown in Table 3. The user's accuracies ranged from 75.0% (class 4: Cattail) to 98.6% (class 5: Common Reed). The producer's accuracies varied from 50.0% (class 4: Cattail) to 97.2% (class 5: Common Reed). Cattail, an invasive species in the Greater Everglades, was the most difficult marsh species to classify, while Common Reed, a giant grass considered as a looming threat to the Greater Everglades, was the easiest to identify for the study site.

The corresponding uncertainty map was also produced from the ensemble analysis, as shown in Figure 3(e). It was difficult to visually identify the difference between the classified maps (Figure 3(a–d)). But the uncertainty map successfully revealed consistency and discrepancy between different classifiers. Most regions (69.2%) showed a full agreement from three classifiers (shown in green), indicating a high confidence for being correctly classified. Some areas (27.8%) (shown in blue) were voted by two classifiers, generating a partial agreement in the classification. A few regions (3.0%) displayed a 'warning sign' (shown in red), indicating a high probability of being

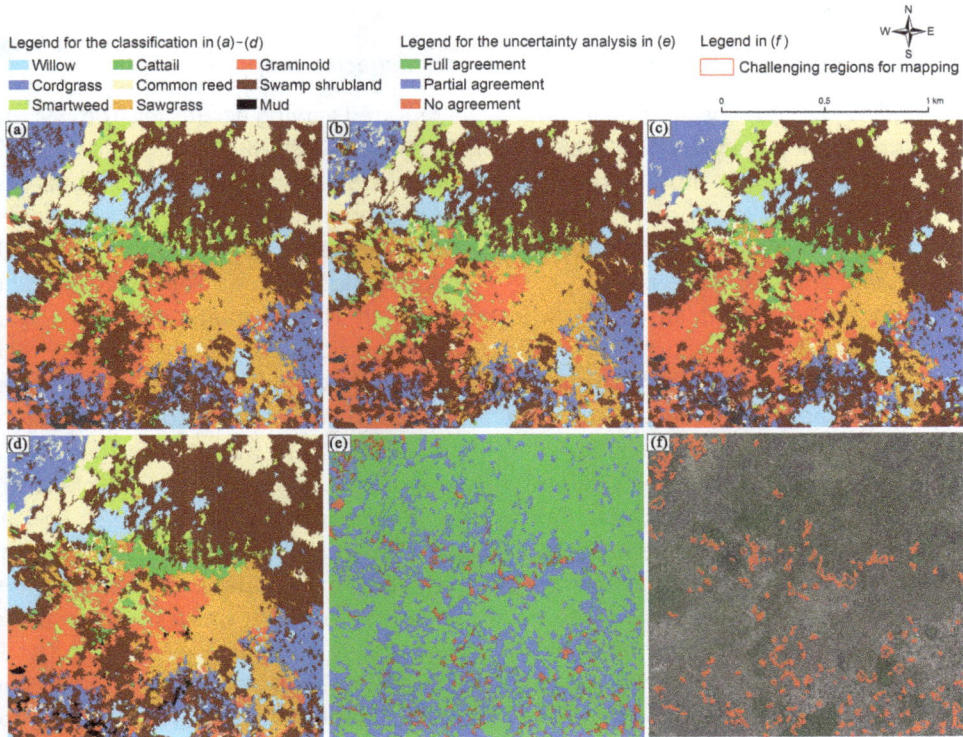

Figure 3. Classified maps from SVM (a), RF (b), ANN (c), and ensemble analysis (d); Uncertainty map from ensemble analysis of SVM, RF, and ANN (e); and identified challenging regions over the natural colour of the aerial photography (f).

Table 3. Error matrix of the ensemble analysis of three classifiers using the fused dataset.

Class	1	2	3	4	5	6	7	8	9	RT	PA (%)
1	52							2		54	96.3
2		41						3		44	93.2
3			40	3			6	7		56	71.4
4	1		7	12				4		24	50.0
5					70	1	1			72	97.2
6						62	4	4		70	88.6
7		2	1			10	56	5		74	75.7
8		5	4	1		3	4	97	1	115	84.4
9					1				26	27	96.3
CT	53	48	52	16	71	76	71	122	27	OA (%): 85.1	
UA (%)	98.1	85.4	76.9	75.0	98.6	81.6	78.9	79.5	96.3	Kappa coefficient: 0.83	

PA (%): producer's accuracy; UA (%): user's accuracy; OA (%): overall accuracy; RT: row total; CT: column total; Classification results are displayed in rows, and the reference data are displayed in columns. The name of each class is displayed in Table 2.

misclassified because there was no agreement from three classifiers for these regions. The regions with the 'warning sign' were further overlaid on the natural colour composite of the aerial photography to highlight them, as shown in Figure 3(f). Overall, the uncertainty map revealed that the final marsh map was robust.

5. Discussion

5.1. *Contribution of texture features in marsh species mapping*

Past studies in marsh species classification have reported varying accuracies from 59% to 99% using optical imagery (Belluco et al. 2006; Sadro, Gastil-Buhl, and Melack 2007; Wang et al. 2007). In our study, a moderate accuracy was obtained if only four bands and NDVI of the aerial photography were used. An averaged OA of 66.1% and kappa coefficient of 0.60 was produced from three classifications. This accuracy is much lower than the accuracy of salt marsh species classification from IKONOS and QuickBird by Belluco et al. (2006), who reported an OA of greater than 95%. Salt marsh plant zonation tends to be much stricter along tidal inundation gradients than the zonation observed among freshwater species, allowing for greater species-level classification accuracy (Carle, Wang, and Sasser 2014). When we added the texture measures in the classification, a higher accuracy was achieved with an averaged OA of 80.3% and kappa coefficient of 0.77 from three classifiers. Our study is comparable to the study from Carle, Wang, and Sasser (2014) who also mapped freshwater marsh species but used WorldView-2 satellite imagery. An OA of 75% and kappa coefficient of 0.71 was reported in their study for classifying eight freshwater marsh species and three other land cover types. Few studies include texture measures in marsh species mapping. Our study revealed an improvement of OA by 14.2% if texture measures were added. This is consistent with the findings from Szantoi et al. (2013, 2015), who also reported the contribution of texture measures derived from fine resolution aerial photography for mapping marsh and wet prairie communities in Everglades National Park. Note that the magnitude of difference in classification accuracies varied across species when texture was incorporated. Some species, such as Cattail and Cordgrass, showed the greatest improvement in accuracy with the addition of texture as opposed to species such as Willow. This could be related to the biophysical properties such as vegetation height and variability within-species canopy architecture which is much more pronounced for Willow compared to Cattail. Further examination showed a high spectral confusion between Cattail and Smartweed, while an inclusion of texture features improved the discrimination for both species.

   Texture has been of great interest in remote sensing for more than three decades with the aim of incorporating spatial features in classification. Warner (2011) presented a detailed review of texture analysis in remote sensing, and concluded that texture measures are useful in image classification but challenging in determining the optimal kernel/window size, an important parameter in texture analysis. Application of OBIA may overcome this challenge by calculating the texture measures in an adaptive window with variable size and shape. OBIA offers the capability to identify regions of varying shapes and sizes in an image, which can be used for subsequent texture extraction (Blaschke 2010). Inclusion of object-based texture measures largely increased the accuracy in marsh species mapping. Other challenges in texture analysis include the specification of texture order, measures, and spectral bands. The setting of these parameters is case- and class-specific (Warner 2011). A range of texture features can be extracted. It is a trade-off to use these features in the mapping procedure (Zhang, Smith, and Fang 2018). First, addition of all these features will increase the data dimensionality, making the classification more complex. Second, an effective feature selection approach needs to be identified in order to determine the most informative features for marsh species classification. Lastly, the effectiveness of these texture features is scale dependent, and selecting an optimal scale for each texture feature is a challenge. In this study, only limited features were used. It is beyond the scope of this study to give a detailed evaluation of each texture feature for marsh species identification.

5.2. *Benefits of a lidar-DEM for marsh species mapping*

Lidar-DEMs have been frequently combined with optical imagery to improve or refine land cover classification in many applications, yet have hardly been utilized for marsh species mapping. A few studies applied a lidar-DEM in a post-classification procedure to character-ize marsh elevations and distribution (Morris et al. 2005; Sadro, Gastil-Buhl, and Melack 2007) or refine marsh species classification into different heights (Hladik, Schalles, and Alber 2013). Here we included object-based lidar-DEM features in the classification rather than a post-classification. An addition of lidar-DEM features achieved an averaged OA of 84.6% and kappa coefficient of 0.83 from three classifiers and ensemble analysis, leading to an increase of 4.3% in OA compared with the application of aerial photography alone. This increase was statistically significant. Topographical features are usually homogeneous within the same marsh, which can help reduce the within-class variability among adjacent objects caused by shadows or gaps, thus increasing the classification accuracy. In addition, hydroperiod is the main factor in determining the spatial distribution of marsh species in the wetlands of Lake Okeechobee. The topography impacts the hydroperiod. Higher elevation means a shorter hydroperiod, while lower elevations have a longer hydroperiod. Inclusion of topographical data was able to improve marsh species classification, as demonstrated in this study. Note that there was a large time gap between the lidar data collection and aerial photography. A simultaneous data collection would be more useful in data fusion. Other lidar features such as elevation and intensity might also help improve marsh species mapping if data were collected simultaneously.

5.3. *Advantages of machine learning classifiers and ensemble analysis*

Past studies commonly applied the traditional classifiers such as Maximum Likelihood and *k*-means to classify multispectral imagery, and Spectral Angle Mapper and linear unmixing to classify hyperspectral imagery for marsh species mapping (e.g. Li, Ustin, and Lay 2005; Belluco et al. 2006; Judd et al. 2007). Application of machine learning classifiers is scarce. Carle, Wang, and Sasser (2014) compared SVM and Maximum Likelihood in freshwater marsh species mapping using WorldView-2 imagery. They found that Maximum Likelihood produced a higher accuracy than SVM. Note that in their study, only spectral bands were used and texture features were not included. When multiple data sources and texture features are combined in the classification, Maximum Likelihood tends to produce a poor result, because this classifier assumes the spectral response of each class displays a Gaussian distribution. An integration of spatial, spectral, and topographical features in the classification can hardly guarantee a Gaussian distribution of the data (Zhang and Xie 2012). Here, we evaluated the performance of three non-parametric machine learning classifiers, SVM, RF and ANN, in marsh species mapping. Three classifiers produced a comparable result but showed diversity in per-class accuracy, which drove us to further evaluate the ensemble analysis in the mapping procedure. The result showed that the ensemble analysis did not improve classification accuracy, and there was no significant difference in the classifications between ensemble analysis, SVM, RF, and ANN. However, ensemble analysis provided an uncertainty map which presented complementary information to the traditional accuracy assessment approaches. The traditional accuracy assessment approaches such as OA, and producer's and user's accuracies only present the quantitative evaluation of the classification and cannot provide any spatial information. The uncertainty map from ensemble analysis can identify regions easy to classify and areas challenging to discriminate. This type of map is useful when there is a desire to minimize the omission or commission errors. It also can be used to guide the post-classification fieldwork (Zhang, Selch, and Cooper 2016). This study, for the first time, provided an uncertainty map of marsh species classification in conjunction with a final classified marsh map. Note that in this study, ensemble analysis was conducted for the outputs of three classifiers and the fused dataset. There is a potential to apply ensemble analysis for the classifications of different datasets. For example, if texture measures or the lidar-DEM did not increase or reduce the classification of a specific class, the discrimination of this class can use the aerial photography alone without texture or topographical features through ensemble analysis. In general, ensemble analysis is useful in marsh species mapping.

5.4. *Limitations of the developed techniques for marsh mapping*

Although a combination of two data sources and four contemporary data processing techniques achieved an encouraging result for marsh species mapping, some limitations still remain. First, there were uncertainties in the application of texture features. Both first-order and second-order textures were extracted and included, but not evaluated separately. Selection of texture orders, different metrics, as well as specification of parameters in calculating texture measures will impact the result (Warner 2011). Second, there are limitations in the application of OBIA in the classification. OBIA requires the inputs of several parameters for generating image objects, and the specifications of these parameters are subjective. A range of methods has been developed to optimize the scale parameter for segmentation such as visual analysis, supervised, and unsupervised approaches (e.g. Drăguţ, Tiede, and Levick 2010; Johnson and Xie 2011). Many of these methods are single-scale, that is, only one scale parameter is used in the segmentation. Single-scale segmentation might lead to both over- and under-segmentation, resulting in uncertainties in the final classification, even though this scale is identified as the optimal scale parameter. A multi-scale segmentation approach, i.e. using multiple scale parameters in the segmentation, may mitigate this problem. But application of a multi-scale segmentation approach is complex. Thus, in this study, the single-scale segmentation was used. Other parameters in the segmentation were set subjectively. The third limitation is the application of machine learning classifiers, which had a good performance in this study. Machine learning methods are data-driven techniques which require adequate training samples to generate ideal results. In addition, each method is sensitive to the input parameters. Setting an optimal parameter is not easy. The selection of these parameters will bring uncertainties in the final classification result. Lastly, note that the developed marsh mapping procedure still needs the input of reference samples which are either from field surveys or manual interpretation methods to calibrate and validate the classification. The amount of available reference data and sampling strategies also impact the final mapping result, and the collection of this reference data depends on traditional methods. Errors of reference data will inevitably propagate into the digital mapping procedure.

6. Conclusions

In this study, we combined two data sources (aerial photography and lidar data), and four data processing techniques (OBIA, machine learning, texture analysis, and ensemble analysis) to map marsh species. Pros and cons of the developed mapping procedure were discussed. Testing of the procedure over a portion of wetlands in Lake Okeechobee produced encouraging results. We found that texture features were useful in marsh species classification. An addition, texture measures increased the OA by 14.2% compared with the application of spectral data alone. Inclusion of a lidar-DEM was also beneficial. An integration of aerial photography and a lidar-DEM achieved an OA of 85.3% and a kappa coefficient of 0.83 using the RF classifier. Three machine learning classifiers had a good performance in classifying aerial photography alone and the fused dataset, and there was no significant difference between these classifiers in the classification. Application of ensemble analysis in marsh species mapping was valuable. Ensemble analysis did not increase the classification accuracy, but it provided a useful uncertainty map which complemented the traditional accuracy assessment approach. This uncertainty map could show regions with a robust classification, and areas where classification errors were most likely to occur. A joint examination of the final classified map and uncertainty map can guide the post-classification and field work.

The developed mapping procedure can be slightly modified to combine more data sources or multi-temporal data for better marsh mapping. Similarly, other classifiers can also be integrated for a more robust ensemble analysis. The classifiers to be combined should have a comparable OA, but different per-class accuracy to make the ensemble analysis useful. More studies are needed to test the developed techniques in other marsh environments, such as tidal marshes, and floating freshwater marsh species to assess the robustness and extensionality. Examination of other data fusion techniques and contribution of different texture features is also valuable. These should be the major directions in our future research. Given the global marsh degradation caused by human activities, climate change, and sea level rise, it is anticipated that this study can benefit regional marsh species mapping in general, and the wetlands of Lake Okeechobee in particular.

Disclosure statement

No potential conflict of interest was reported by the authors.

References

Anguita, D., L. Ghelardoni, A. Ghio, L. Oneto, and S. Ridella. 2012. "The 'K' in K-Fold Cross Validation." In *ESANN 2012 Proceedings, European Symposium on Artificial Neural Networks, Computational Intelligence and Machine Learning*, Bruges, Belgium, April 25–27.

Belgiu, M., and L. Drăguţ. 2016. "Random Forest in Remote Sensing: A Review of Applications and Future Directions." *ISPRS Journal of Photogrammetry and Remote Sensing* 114: 24–31. doi:10.1016/j.isprsjprs.2016.01.011.

Belluco, E., M. Camuffo, S. Ferrari, L. Modenese, S. Silvestri, A. Marani, and M. Marani. 2006. "Mapping Salt-Marsh Vegetation by Multispectral and Hyperspectral Remote Sensing." *Remote Sensing of Environment* 105: 54–67. doi:10.1016/j.rse.2006.06.006.

Blaschke, T. 2010. "Object Based Image Analysis for Remote Sensing." *ISPRS Journal of Photogrammetry and Remote Sensing* 65: 2–16. doi:10.1016/j.isprsjprs.2009.06.004.

Breiman, L. 2001. "Random Forests." *Machine Learning* 45: 5–32. doi:10.1023/A:1010933404324.

Carle, M. V., L. Wang, and C. E. Sasser. 2014. "Mapping Freshwater Marsh Species Distributions Using WorldView-2 High-Resolution Multispectral Satellite Imagery." *International Journal of Remote Sensing* 35: 4698–4716. doi:10.1080/01431161.2014.919685.

CERP. 2018. *Comprehensive Everglades Restoration Plan.* Accessed 20 March 2018. https://www.evergladesrestoration.gov/.

Doren, R. F., K. Rutchey, and R. Welch. 1999. "The Everglades: A Perspective on the Requirements and Applications for Vegetation Map and Database Products." *Photogrammetric Engineering & Remote Sensing* 65: 155–161.

Drăguţ, L., D. Tiede, and S. R. Levick. 2010. "ESP: A Tool to Estimate Scale Parameter for Multiresolution Image Segmentation of Remotely Sensed Data." *International Journal of Geographical Information Science* 24: 859–871. doi:10.1080/13658810903174803.

Du, P., J. Xia, W. Zhang, K. Tan, Y. Liu, and S. Liu. 2012. "Multiple Classifier System for Remote Sensing Image Classification: A Review." *Sensors* 12: 4764–4792. doi:10.3390/s120404764.

Foody, G. M. 2004. "Thematic Map Comparison, Evaluating the Statistical Significance of Differences in Classification Accuracy." *Photogrammetric Engineering & Remote Sensing* 70: 627–633. doi:10.14358/PERS.70.5.627.

Gilmore, M. S., E. H. Wilson, N. Barrett, D. L. Civco, S. Prisloe, J. D. Hurd, and C. Chadwick. 2008. "Integrating Multi-Temporal Spectral and Structural Information to Map Wetland Vegetation in a Lower Connecticut River Tidal Marsh." *Remote Sensing of Environment* 112: 4048–4060. doi:10.1016/j.rse.2008.05.020.

Gómez-Chova, L., D. Tuia, G. Moser, and G. Camps-Valls. 2015. "Multimodal Classification of Remote Sensing Images: A Review and Future Directions." *Proceedings of the IEEE* 103: 1560–1584. doi:10.1109/JPROC.2015.2449668.

Grybas, H., L. Melendy, and R. G. Congalton. 2017. "A Comparison of Unsupervised Segmentation Parameter Optimization Approaches Using Moderate- and High-Resolution Imagery." *GIScience & Remote Sensing* 54: 515–533. doi:10.1080/15481603.2017.1287238.

Hall, M., E. Frank, G. Holmes, B. Pfahringer, P. Reutemann, and I. H. Witten. 2009. "The WEKA Data Mining Software: An Update." *ACM SIGKDD Explorations Newsletter* 11: 10–18. doi:10.1145/1656274.

Hall-Beyer, M. 2017. "GLCM Texture: A Tutorial v. 3.0." March 2017.

Havens, K. E., and D. E. Gawlik. 2005. "Lake Okeechobee Conceptual Ecological Model." *Wetlands* 25: 908–925. doi:10.1672/0277-5212(2005)025[0908:LOCEM]2.0.CO;2.

Hladik, C., J. Schalles, and M. Alber. 2013. "Salt Marsh Elevation and Habitat Mapping Using Hyperspectral and LIDAR Data." *Remote Sensing of Environment* 139: 318–330. doi:10.1016/j.rse.2013.08.003.

Johnson, B., and Z. Xie. 2011. "Unsupervised Image Segmentation Evaluation and Refinement Using a Multi-Scale Approach." *ISPRS Journal of Photogrammetry and Remote Sensing* 66: 473–483. doi:10.1016/j.isprsjprs.2011.02.006.

Judd, C., S. Steinberg, F. Shaughnessy, and G. Crawford. 2007. "Mapping Salt Marsh Vegetation Using Aerial Hyperspectral Imagery and Linear Unmixing in Humboldt Bay, California." *Wetlands* 27: 1144–1152. doi:10.1672/0277-5212(2007)27[1144:MSMVUA]2.0.CO;2.

Kumar, L., and P. Sinha. 2014. "Mapping Salt-Marsh Land-Cover Vegetation Using High-Spatial and Hyperspectral Satellite Data to Assist Wetland Inventory." *GIScience & Remote Sensing* 51: 483–497. doi:10.1080/15481603.2014.947838.

Lake Okeechobee Watershed Restoration Project. 2018. Accessed 20 March 2018. https://www.sfwmd.gov/our-work/cerp-project-planning/lowrp.

Landis, J., and G. G. Koch. 1977. "The Measurement of Observer Agreement for Categorical Data." *Biometrics* 33: 159–174. doi:10.2307/2529310.

Li, L., S. L. Ustin, and M. Lay. 2005. "Application of Multiple Endmember Spectral Mixture Analysis (MESMA) to AVIRIS Imagery for Coastal Salt Marsh Mapping: A Case Study in China Camp, CA, USA." *International Journal of Remote Sensing* 26: 5193–5207. doi:10.1080/01431160500218911.

Mas, J. F., and J. J. Flores. 2008. "The Application of Artificial Neural Networks to the Analysis of Remotely Sensed Data." *International Journal of Remote Sensing* 29: 617–663. doi:10.1080/01431160701352154.

Mishra, D. R., S. Ghosh, C. Hladik, J. L. O'Connell, and H. J. Cho. 2015. "Wetland Mapping Methods and Techniques Using Multi-Sensor, Multi-Resolution Remote Sensing: Successes and Challenges." In *Remote Sensing Handbook*, edited by P. S. Thenkabail, 191–226. Vol. III. ISBN: 13: 978-1-4822-1972-6, CAT# K22131. Boca Raton: CRC Press, Taylor and Francis.

Morris, J. T., D. Porter, M. Neet, P. A. Noble, L. Schmidt, L. A. Lapine, and J. R. Jensen. 2005. "Integrating LIDAR Elevation Data, Multi-Spectral Imagery and Neural Network Modelling for Marsh Characterization." *International Journal of Remote Sensing* 26: 5221–5234. doi:10.1080/01431160500219018.

Mountrakis, G., J. Im, and C. Ogole. 2011. "Support Vector Machines in Remote Sensing: A Review." *ISPRS Journal of Photogrammetry and Remote Sensing* 66: 247–259. doi:10.1016/j.isprsjprs.2010.11.001.

Rutchey, K., T. N. Schall, R. F. Doren, A. Atkinson, M. S. Ross, D. T. Jones, M. Madden, et al. 2006. *Vegetation Classification for South Florida Natural Areas*, 142. Saint Petersburg: United States Geological Survey. Open-File Report 2006-1240.

Sadro, S., M. Gastil-Buhl, and J. Melack. 2007. "Characterizing Patterns of Plant Distribution in a Southern California Salt Marsh Using Remotely Sensed Topographic and Hyperspectral Data and Local Tidal Fluctuations." *Remote Sensing of Environment* 110: 226–239. doi:10.1016/j.rse.2007.02.024.

Sun, C., Y. Liu, S. Zhao, M. Zhou, Y. Yang, and F. Li. 2016. "Classification Mapping and Species Identification of Salt Marshes Based on a Short-Time Interval NDVI Time-Series from HJ-1 Optical Imagery." *International Journal of Applied Earth Observation and Geoinformation* 45: 27–41. doi:10.1016/j.jag.2015.10.008.

Szantoi, Z., F. Escobedo, A. Abd-Elrahman, L. Pearlstine, B. Dewitt, and S. Smith. 2015. "Classifying Spatially Heterogeneous Wetland Communities Using Machine Learning Algorithms and Spectral and Textural Features." *Environmental Monitoring and Assessment* 187: 262. doi:10.1007/s10661-015-4426-5.

Szantoi, Z., F. Escobedo, A. Abd-Elrahman, S. Smith, and L. Pearlstine. 2013. "Analyzing Fine-Scale Wetland Composition Using High Resolution Imagery and Texture Features." *International Journal of Applied Earth Observation and Geoinformation* 23: 204–212. doi:10.1016/j.jag.2013.01.003.

Trimble. 2014. *eCognition Developer 9.0.1 Reference Book*. Munich, Germany: Trimble Germany GmbH.

Vapnik, V. N. 1995. *The Nature of Statistical Learning Theory*. New York: Springer-Verlag.

Wang, C., M. Menenti, M. Stoll, E. Bellucob, and M. Marani. 2007. "Mapping Mixed Vegetation Communities in Salt Marshes Using Airborne Spectral Data." *Remote Sensing of Environment* 107: 559–570. doi:10.1016/j.rse.2006.10.007.

Warner, T. 2011. "Kernel-Based Texture in Remote Sensing Image Classification." *Geography Compass* 5: 781–798. doi:10.1111/geco.2011.5.issue-10.

Zhang, C. 2014. "Combining Hyperspectral and LiDAR Data for Vegetation Mapping in the Florida Everglades." *Photogrammetric Engineering & Remote Sensing* 80: 733–743. doi:10.14358/PERS.80.8.733.

Zhang, C. 2015. "Applying Data Fusion Techniques for Benthic Habitat Mapping and Monitoring in a Coral Reef Ecosystem." *ISPRS Journal of Photogrammetry and Remote Sensing* 104: 213–223. doi:10.1016/j.isprsjprs.2014.06.005.

Zhang, C., D. Selch, and H. Cooper. 2016. "A Framework to Combine Three Remotely Sensed Data Sources for Vegetation Mapping in the Central Florida Everglades." *Wetlands* 36: 201–213. doi:10.1007/s13157-015-0730-7.

Zhang, C., M. Smith, and C. Fang. 2018. "Evaluation of Goddard's LiDAR, Hyperspectral, and Thermal Data Products for Mapping Urban Land-Cover Types." *GIScience & Remote Sensing* 55: 90–109. doi:10.1080/15481603.2017.1364837.

Zhang, C., and Z. Xie. 2012. "Combining Object-Based Texture Measures with a Neural Network for Vegetation Mapping in the Everglades from Hyperspectral Imagery." *Remote Sensing of Environment* 124: 310–320. doi:10.1016/j.rse.2012.05.015.

Zhang, C., and Z. Xie. 2013. "Object-Based Vegetation Mapping in the Kissimmee River Watershed Using HyMAP Data and Machine Learning Techniques." *Wetlands* 33: 233–244. doi:10.1007/s13157-012-0373-x.

Zhang, C., and Z. Xie. 2014. "Data Fusion and Classifier Ensemble Techniques for Vegetation Mapping in the Coastal Everglades." *Geocarto International* 29: 228–243. doi:10.1080/10106049.2012.756940.

Zhang, C., Z. Xie, and D. Selch. 2013. "Fusing LiDAR and Digital Aerial Photography for Object-Based Forest Mapping in the Florida Everglades." *GIScience & Remote Sensing* 50: 562–573.

Satellite-based salt marsh elevation, vegetation height, and species composition mapping using the superspectral WorldView-3 imagery

Antoine Collin, Natasha Lambert and Samuel Etienne

ABSTRACT

Very high resolution (VHR) space-borne data are needed to finely and continuously map salt marshes. The WorldView-3 (WV-3) sensor leverages one panchromatic, eight optical, and eight shortwave infrared (SWIR) bands at 0.31, 1.24, and 7.5 m pixel size, respectively. Although eight optical bands have been previously pansharpened, no attempt to use the 16-band superspectral data set at VHR (0.31 m) has been yet reviewed. Here, we propose to reliably pan-sharpen the 16 WV-3 predictors so as to model (artificial neural network, ANN) salt marsh elevation and vegetation height and classify species composition at VHR using calibration/validation handheld vegetation height, airborne lidar elevation, and drone blue-green-red (BGR) responses. Three models have been created over a megatidal bay (Beaussais Bay, Brittany, France) provided with mud flats, salt marshes, and polders. VHR-screened WV-3 bands very satisfactorily predicted salt marsh elevation and vegetation height responses ($r = 0.86$, $R^2 = 0.71$, root mean square error (RMSE) = 0.33 m and $r = 0.88$, $R^2 = 0.77$, RMSE = 5.72 cm, respectively). The WV-3 superspectral data set outperformed the eight-band multispectral and four-band traditional data sets to classify 15 salt marsh habitats (OA = 95.47, 82.33, and 69.27%, respectively). Adding WV-3-based salt marsh elevation and vegetation height augmented the 15-class classification of the superspectral and traditional data sets (OA = 97.60 and 77.47%, respectively), but not for the multispectral one (OA = 81.93%).

1. Introduction

Salt marsh ecosystems offer a wide spectrum of goods and services, such as fish, amphibian, and bird ecological niches, herb supply, coastal protection, carbon sequestration, or aesthetic landscapes, among others (Barbier et al. 2011). These eco- service hotspots result from complex interactions among tidal, hydrological, geomorphic, edaphic, ecological, and anthropogenic drivers (Allen 2000; Bertness, Ewanchuk, and Silliman 2002; Pennings, Grant, and Bertness 2005). These coastal ecosystems are strongly threatened by eutrophication (Deegan et al. 2012), hard engineering

(Chapman and Underwood 2011), and agricultural conversion (Gedan, Silliman, and Bertness 2009). In the context of adaptation to accelerated sea-level rising, a fine-scale monitoring of salt marsh composition, structure, and elevation is needed for optimizing the management and spatial planning of their associated ecosystem services. Salt marsh mapping is traditionally based on either passive multispectral/hyperspectral (Belluco et al. 2006) or active light detection and ranging (lidar, Collin, Long, and Archambault 2010) remote-sensing techniques. As reliable proxies for composition and elevation, hyperspectral and lidar data have been combined to successfully improve salt marsh classification (Hladik, Schalles, and Alber 2013). Considering spectral signatures of the water, soil, and vegetation communities with their elevation contributes to better delineation of the zonation of the coastal communities and environments. However, hyperspectral and lidar data are collected by an airborne campaign, which is extremely costly in budget and preparation. Moreover, the accuracy of lidar measurements is too coarse to quantify the structure, i.e. height, of the salt marsh herbaceous vegetation.

In 2009, the space-borne passive WorldView-2 (WV-2) sensor was launched to capture one panchromatic band at 0.46 m and eight optical bands at 1.84 m, enabling a cost-efficient identification and mapping of natural vegetation on a coastal site (Rapinel et al. 2014) and an estimation of high-density biomass for wetland vegetation (Mutanga, Adam, and Cho 2012). In 2014, the WorldView-3 (WV-3) augmented the capabilities of its predecessor by increasing the panchromatic and multispectral spatial resolutions of 0.15 m and 0.6 m, respectively, besides adding eight mid-infrared, called shortwave infrared (SWIR), bands at 3.7 m (resampled at 7.5 m) pixel size. Although its increased spatial resolution produced better bathymetry extraction (Collin, Etienne, and Feunteun 2017), and its augmented number of spectral bands permitted identification and mapping of some key minerals (Kruse and Perry 2013), the WV-3 sensor has never been used for salt marsh purposes. Because the spatial resolution affects salt marsh classification accuracy much more importantly than spectral resolution (Belluco et al. 2006) and because the SWIR spectral range is strongly correlated with water content (Adam, Mutanga, and Rugege 2010), the use of the 16 bands at very high resolution (VHR), provided with the WV-3 sensor, holds great promise to monitor the main factors of salt marshes.

Towards a fine-scale but low-cost mapping of salt marsh ecology, we propose, in this article, to produce the first VHR mapping of salt marsh species composition, height, and elevation (i.e. altitude) based on space-borne predictors calibrated with ground- and air-truth. The superspectral WV-3 sensor, providing 16 optical and SWIR bands at 0.31 m in a pan-sharpening mode, will be trained with ground measurements of vegetation height, airborne lidar soundings of ground elevation, and airborne drone blue-green-red (BGR) spectral signatures of salt marsh components. Artificial neural networks (ANNs) will model the relationships among salt marsh elevation, vegetation height, and species composition responses and WV-3 spectral predictors. Our study was carried out in the southern Channel Sea, over the eastern half of Beaussais Bay (Brittany, France, Figure 1), a bay provided with mud flats, salt marshes, and polders (low-lying tracts of land surrounded by dikes) and submitted to very high tides (13 m). Following the WV-3 imagery preprocessing, the ground and airborne observations, and the modelling methodology, we will spatially model the salt marsh vegetation height, elevation, and species composition at VHR over the entire area. Spectrally based predictions will then be discussed and brought into broader perspectives.

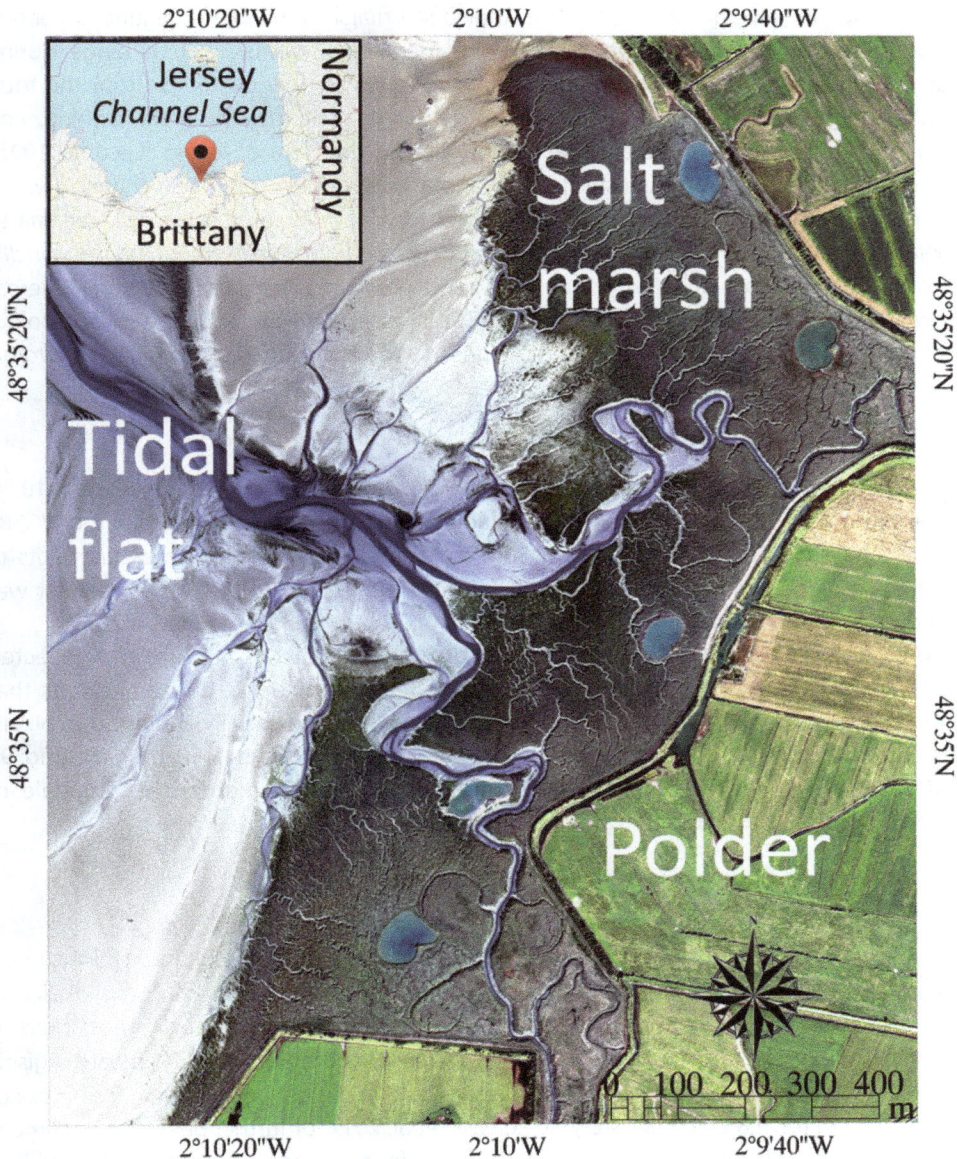

Figure 1. Natural-coloured imagery (WorldView-3 blue, green, and red bands) of the study area in Beaussais Bay (Brittany, France). This imagery was collected on 1 October 2015 and provides spectral information at 0.31 m pixel size (4318 × 5573 pixels).

2. Materials and methods

2.1. *Study site*

The study site is located along the Emerald Coast bordering the Channel Sea (48°35′N, 2°10′W) in Brittany (France, Figure 1), characterized by a temperate climate under oceanic influences. The coastal site extends over 2.11 km2 and tops at 9.62 m (IGN69 reference is the mean sea level) for a mean value of 4.2 ± 1.3 m. The studied salt marsh stems from the junction of two coastal rivers, Floubalay and Drouet, with the megatidal seashore (13 m range during equinoxes). Landward areas were converted, across the last few centuries (from the fourteenth to eighteenth centuries), into market gardening, salt making, grain farming, and ranching through multi-temporal reclamation works in the form of levees (Leconte 1991). This heterogeneous mosaic is constituted by various waterbodies (salt marsh channel, waterfowl pond, polder channel), sediment types (wet, dry, and levee), macroalgae, salt marsh vegetation (*Salicornia europea*, *Halimione portulacoides*, *Triglochin maritima*, *Puccinellia maritima*, *Festuca rubra* and *Elymus athericus*), and polder grazing land (Table 1). Whereas the whole site is classified under the Habitats Directive in the Natura 2000 network, polder areas are the property of the French Coastal Conservation Authority (*Conservatoire du littoral*).

2.2. Handheld vegetation height response and photo-quadrats

A handheld fieldwork was carried out in February 2017 using a decimetre scale and a 10 Hz Global Navigation Satellite System (GNSS, including GPS and GLONASS) device in the WGS84 geodetic system tailored with the UTM 30N projection. An array of 96 measurements of vegetation height was evenly distributed across four transects (Figure 2). The data set was then divided into 64 training and 32 validation samples (white circles in Figure 2).

Contiguously, a series of 90 geolocated photo-quadrats (0.5 m × 0.5 m) was collected across four transects. Each of the 15 classes was featured with six photo-quadrats that were statistically grown to the two closest (spectrally and spatially) neighbour pixels, reaching 18 sampling points per class. This data set was split to 180 training and 90 validation points, dedicated to testing (through a confusion matrix) the rationale of using the drone as air-truth.

2.3. Airborne drone visible response

An airborne BGR camera mounted on a consumer-grade drone (DJI Phantom 3 Pro) was deployed on 7 February 2017 over the Beaussais salt marsh. A series of 442 nadiral BGR photographs, *XYZ* geolocated, were aligned into four transects then orthomosaicked in WGS84 UTM 30 (see BGR mosaics in Figure 2) using photogrammetry software Agisoft Photoscan (http://www.agisoft.ru). Mean horizontal accuracies derived from the four mosaics reached 0.02 m (i.e. 1.85 cm pixel size) (see Casella et al. 2017 for further details). A maximum likelihood pre-classification of the 15 classes over the four mosaics was performed, providing extremely satisfactory results (Table 2, overall accuracy, OA = 91%, Congalton and Green 2009). Such a spatial resolution applied to natural-coloured spectral data set was thereafter considered as detailed enough to be used as air-truth (Collin et al. 2018). A data set of 4500 sampling points was classified into 15 salt marsh classes (Figure 2 and Table 1) and then subdivided into 3000 training and 1500 validation sub-data sets.

2.4. Airborne lidar elevation response

An airborne lidar campaign was conducted in 2011 by the French National Geographic Institute in the frame of the national program Litto3D©. The lidar system provided a minimum sounding density of one point per square metre, with vertical and horizontal

Table 1. Description of the 15 classes identified on airborne drone blue-green-red imagery across the study site (0.02 m spatial resolution).

Class name			Class Description	Drone-based Photograph
Abiotic	**Sediment**	Wet sediment	Clay, silt or sand filled with water on tidal flat	
		Dry sediment	Clay, silt, sand mostly desiccated on tidal flat	
		Levee sediment	Clay or silt in front of levee on upper marsh	
	Water	Salt marsh creek	River-driven network in intertidal area	
		Waterfowl pond	Small lagoon dug by waterfowl hunters	
		Polder channnel	River-driven network in reclaimed area	
Biotic	**Tidal flat**	Macroalgae	Bio-complex of red, brown and green algae	
	Salt marsh	*Salicornia*	Glasswort *S. europea* on lower marsh	
		Triglochin	Arrowgrass *T. maritima* on lower-middle marsh	
		Halimione	Orach *H. portulacoides* on middle-upper marsh	
		Puccinellia	Sweetgrass/manna grass *P. maritima* on upper marsh	
		Festuca	Fescue *F. rubra* on upper marsh	
		Elymus	Quackgrass *E. athericus* on levee	
	Polder	Mown grazing land	Mown permanent or temporary pasture land	
		Standing grazing land	Standing permanent or temporary pasture land	

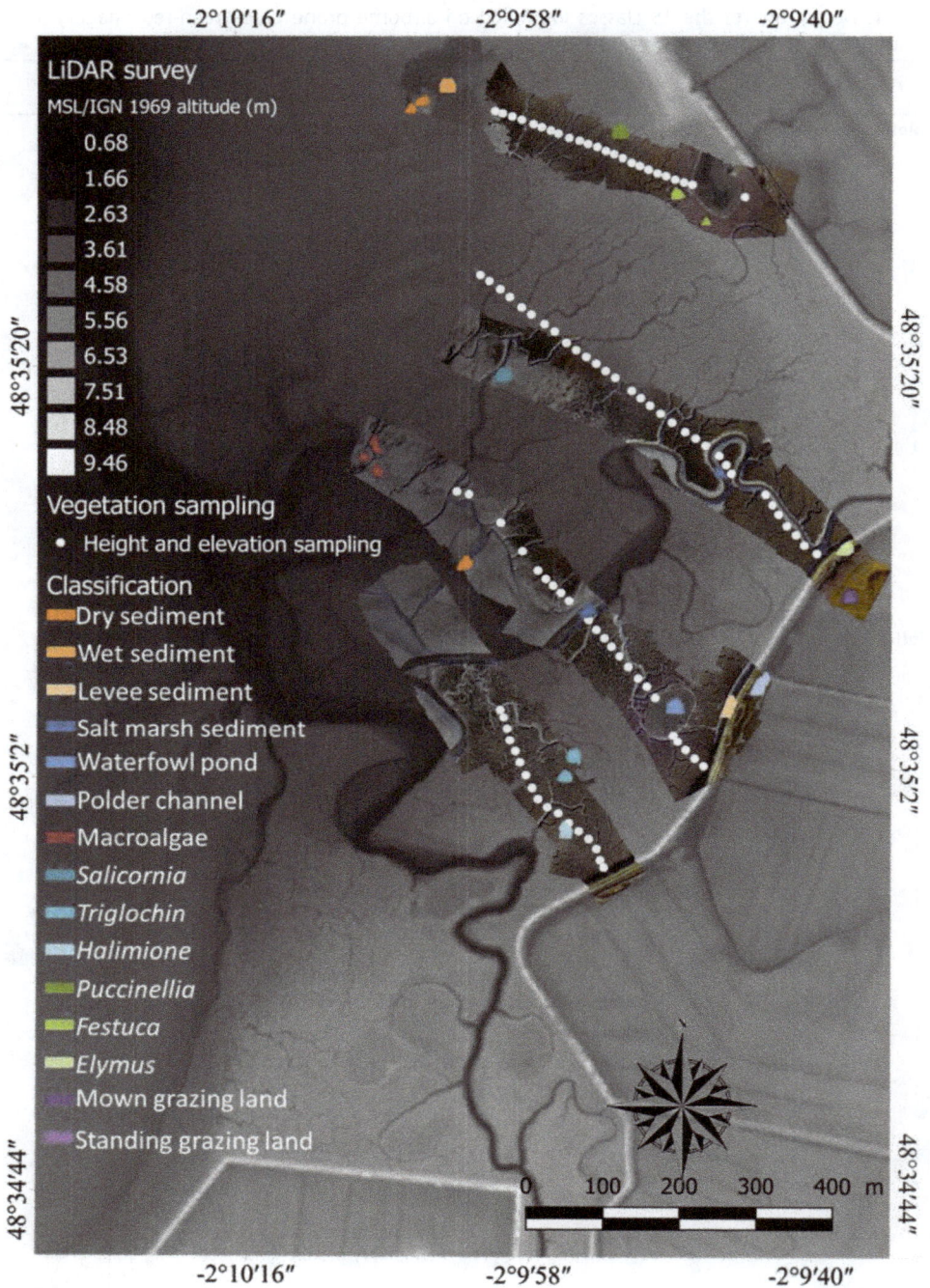

Figure 2. Map of air- and ground-truth data sets combining airborne lidar digital elevation model (greyscale) at 1 m spatial resolution (1420 × 1768 pixels), four airborne drone blue-green-red orthomosaics at 0.02 m pixel size, from which are retrieved 4500 sampling pixels classified into 15 classes (coloured triangles) and 96 sampling points (white circles), where vegetation height and elevation were measured.

Table 2. Confusion matrix synthesizing the classification of 15 selected classes based on the airborne drone blue-green-red imagery.

	W. s.	D. s.	L. s.	S. m. c.	W. p.	P. c.	M.	S.	T.	H.	P.	F.	E.	M. g. l.	S. g. l.	TOTAL
W. s.	6															6
D. s.		5	1					1								7
L. s.			6													6
S. m. c.				6												6
W. p.					6											6
P. c.					1	5										6
M.							6									6
S.								5								5
T.									4							4
H.									1	6						7
P.									1		5				1	7
F.									1			6				7
E.													5			5
M. g. l.														6		6
S. g. l.														1	5	6
TOTAL	6	5	6	6	7	6	6	6	7	6	5	6	6	6	6	90

W. s., wet sediment; D. s., dry sediment; L. s., levee sediment; S. m. c., salt marsh creek; W. p., waterfowl pond; P. c., polder channel; M., macroalgae; S.: *Salicornia*; T., *Triglochin*; H., *Halimione*; P., *Puccinellia*; F., *Festuca*; E., *Elymus*; M. g. l., mown grazing land; S. g. l.,standing grazing land.

accuracies of 0.2 and 0.5 m, respectively (Litto3D© 2015). The topographic lidar pulses an electromagnetic radiation (1064 nm wavelength, i.e. near-infrared, NIR) from the aircraft and records its travel time in air and water by means of a waveform (Collin, Long, and Archambault 2010). Lidar elevation is computed on board for each sounding by converting the time between the aircraft and the ground surface

NIR echo into distance (knowing the light speed into air). Because each lidar elevation is associated with the sounding geolocation, as a by-product of the joint GNSS and inertial measurement unit, a digital elevation model (DEM, Figure 2) is computed using the Kriging method applied to lidar sounding clouds. The DEM was geographically referenced to WGS84 UTM 30N and altimetrically referenced to the mean sea level (IGN69).

2.5. *Space-borne WV-3 superspectral predictors*

The WV-3 imagery was collected on 1 October 2015 at 11h34min45sec (UTC) over Beaussais Bay. Beyond the traditional Blue-Green-Red-InfraRed (BGRIR) gamut, the WV-3 imagery is composed of one panchromatic, eight optical, and eight SWIR bands, provided with 0.31, 1.24, and 3.7 m spatial resolutions, respectively. Each spectral band (out of 16) has a specific spectral sensitivity that could be synergistically combined together for salt marsh purposes (Figure 3). While the panchromatic and optical bands were released as such, the SWIR bands were resampled at 7.5 m spatial resolution.

Given the ortho-rectification (WGS84 UTM 30N) implemented by the DigitalGlobe foundation, no geometric correction was applied. Prior to pan-sharpening, radiometric correction was applied, by, first, converting digital numbers (DNs, 11 bits for the optical and 14 bits for the SWIR bands, Figure 4(a)) into top-of-atmosphere (TOA, Figure 4(b)) radiance using calibration gains and offsets (.imd file), and then into down-of-atmosphere (DOA, Figure 4(c)) radiance and reflectance correcting for atmosphere interactions and sun irradiance (MODTRAN®-based FLAASH® module in geospatial ENVI software, Harris Geospatial

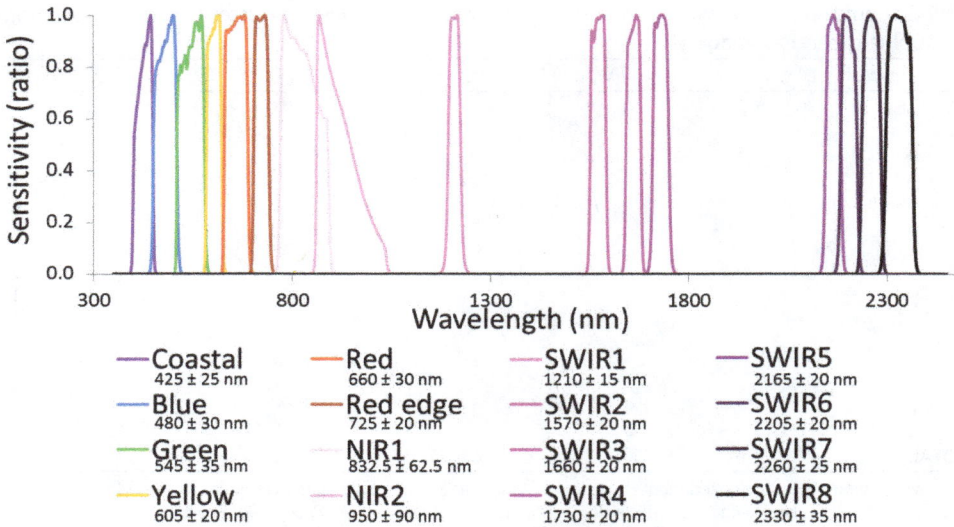

Figure 3. Line plot of the 16 spectral sensitivities of the WorldView-3 space-borne sensor (wavelength centre ± width/2 nm).

Solutions, Broomfield CO, USA). The spatial resolutions of both optical and SWIR imageries can be scaled up to this of panchromatic using the pan-sharpening procedure (pixel-scale fusion technique). Five steps are needed: (1) up-scaling 1.24 m optical and 7.5 m SWIR imageries to 0.31 m, (2) applying the Gram–Schmidt forward transformation on both up-scaled imageries to express the original pixel reflectance into a transformed spectral space, (3) matching the transformed multispectral reflectance to the transformed panchromatic reflectance, (4) substituting the transformed panchromatic band for the transformed multi-spectral bands, and (5) applying the Gram–Schmidt reverse to the substituted multispectral bands to the original spectral space (see Collin, Archambault, and Planes 2013 for further details). The first eight optical and last eight SWIR bands were processed separately, and then merged together, to obtain 16 spectral bands at 0.31 m. The substitution technique was based on the WV-3 sensor sensitivity ratio (filter function) and a bilinear resampling. Regarding the VHR (original) panchromatic band ranges from blue to NIR-1, a control of the DOA reflectance stability between the original (Figure 4(c)) and pansharpened (Figure 4(d)) data sets was driven across the 15 studied classes (i.e. 4500 pixels, 300 pixels each class). Similar spectral patterns were conspicuous for both data sets, which contributed to confirm the validity of the WV-3 VHR (0.31 m) superspectral (16-band) data set.

2.6. *ANN modelling*

The ANN learner was selected as the regression technique for modelling the salt marsh elevation, vegetation height, and species composition. The ANN has the capability of building an internal representation of an image pattern. The dynamic adjustment of the network uses a proposed input (training data set) and an aimed output to initiate feedback into the ANN, and thus the back-propagation term. The greatest advantages of this process are the absence of *a priori* programming, the tolerance of missing pixel

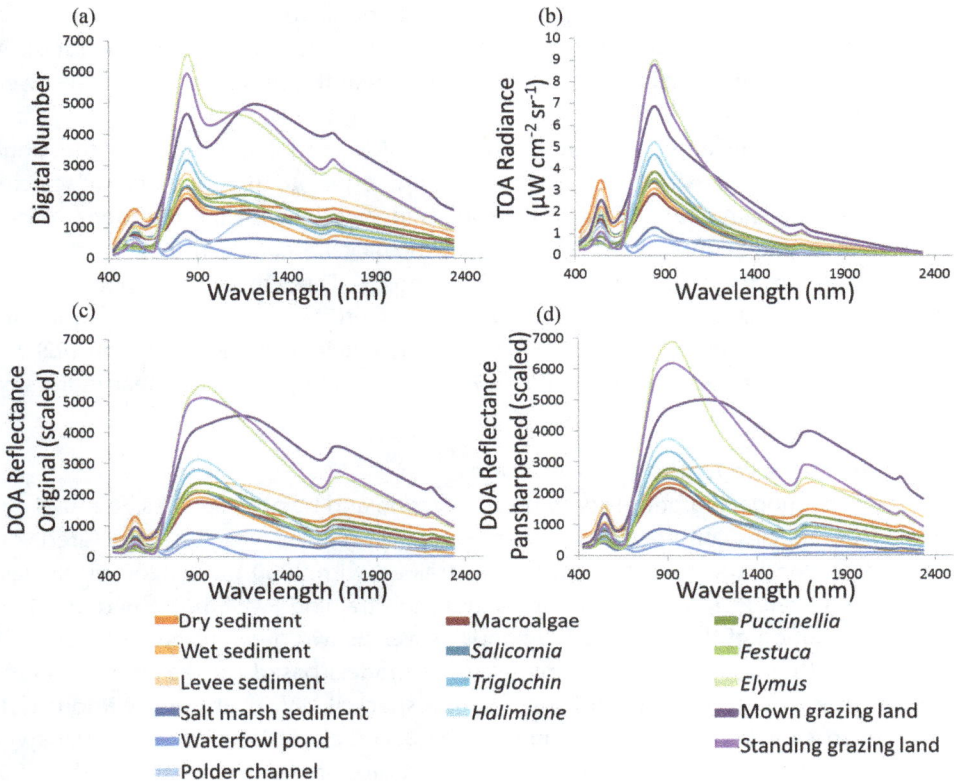

Figure 4. Spectral signatures of the 15 classes (a) studied for radiometrically uncorrected digital numbers: 11 and 14 bits for the first eight and last eight WorldView-3 spectral bands, respectively, (b) converted into top-of-atmosphere (TOA) radiance, (c) corrected for atmosphere interactions and sun irradiance, thus converted into down-of-atmosphere (DOA) reflectance, and (d) augmented as pansharpened DOA reflectance. Both DOA reflectances were scaled between 0 and 10,000 to deal with integers.

values, and the integration of noise in imagery. These features lead to classifications provided with relatively homogenous regions, sharp transition boundaries, and continuous connected features, which are characteristically speckled on conventional supervised classifications (Collin, Etienne, and Feunteun 2017).

The ANN builds nonlinear models, h, by minimizing least squares using a one- layer perceptron feed-forward technique, approximating the i (here, $i = 3$) training responses with a constant, k, and the WV-3 VHR spectral predictors, X, through weighted functions, w_i, adjusting neurons, n_i, that are based on the hyperbolic tangent activation function (Heermann and Khazenie 1992):

$$h(X) = k\left(\sum_i w_i n_i(X)\right) \tag{1}$$

2.6.1. *Salt marsh elevation and vegetation height*
Prior to ANN modelling, a stepwise ordinary least squares (linear) regression linked the 96 elevation and height ground samples with the corresponding space-borne spectral

pixels. This pre-regression enabled the best spectral predictors to be selected based on the absolute value of the predictor estimate divided by its standard error (t ratio). All predictors provided with an absolute t ratio greater than the 0.05 significance level were kept for further ANN modelling.

For both elevation and height responses, an ANN model was built on the single hidden layer provided with a number of neurons equal to the number of selected predictors − 1, so that the number of neurons is in synergy with the number of inputs (Collin, Etienne, and Feunteun 2017). Trained by the 64 calibration samples, the ANN was validated by the remaining 32 validation samples. The consistency between the observed and predicted elevation and height was quantified using the Pearson product-moment correlation coefficient (r), as well as the coefficient of determination ($R2$) and the associated root mean square error (RMSE) corresponding to a final linear regression.

2.6.2. *Salt marsh vegetation classification*
Along with the modelling of the continuous elevation and height variables, the ANN also modelled the species composition response categorized into 15 classes. Structured with the single hidden layer, the ANN model was trained with 3000 pixels (200 pixels each class) derived from BGR air drone samples, until the training RMSE bottomed at 0.1.

The contribution of the WV-3 added bands as well as two previous models (see 2.6.1) was quantified by building and comparing ANN models based on the four traditional BGRIR, the eight multispectral, and the 16 superspectral WV-3 with and without both WV-3-derived elevation and height models. The accuracy of the six classifications was estimated through the OA resulting from the confusion matrix based on the 1500 validation pixels (100 pixels each class).

3. Results

3.1. *Salt marsh vegetation height*

The screening of the WV-3 spectral explanators of the vegetation height showed better performance (absolute t ratio) from SWIR rather than optical wavebands (see silver asterisks in Figure 5(a)). Only the SWIR-3 band did not outreach the 0.05 significance threshold (t ratio = 0.23) in the SWIR gamut, whereas only the blue band overcame the threshold (t ratio = 2.01) in the optical spectrum. The best predictor was the SWIR-1 band (t ratio = 5.36). By combining the selected eight best predictors, the ANN model, based on seven neurons within the single hidden layer, was trained with the 64 calibration pixels. The resulting seven-neuroned model was then tested against the 32 validation pixels, resulting in a very good agreement (r = 0.88, $R2$ = 0.77, and RMSE = 5.72cm, Figure 5(c)).

The ANN formula was then applied to each pixel of the stacked blue, SWIR-1, SWIR-2, SWIR-4, SWIR-5, SWIR-6, SWIR-7, and SWIR-8 bands in order to continuously map the salt marsh vegetation height (digital vegetation height model, DVHM) provided with 0.31 m spatial resolution (Figure 6(a)). The visual rendering indicates an intuitive mean increase of the vegetation height from lower to higher salt marsh.

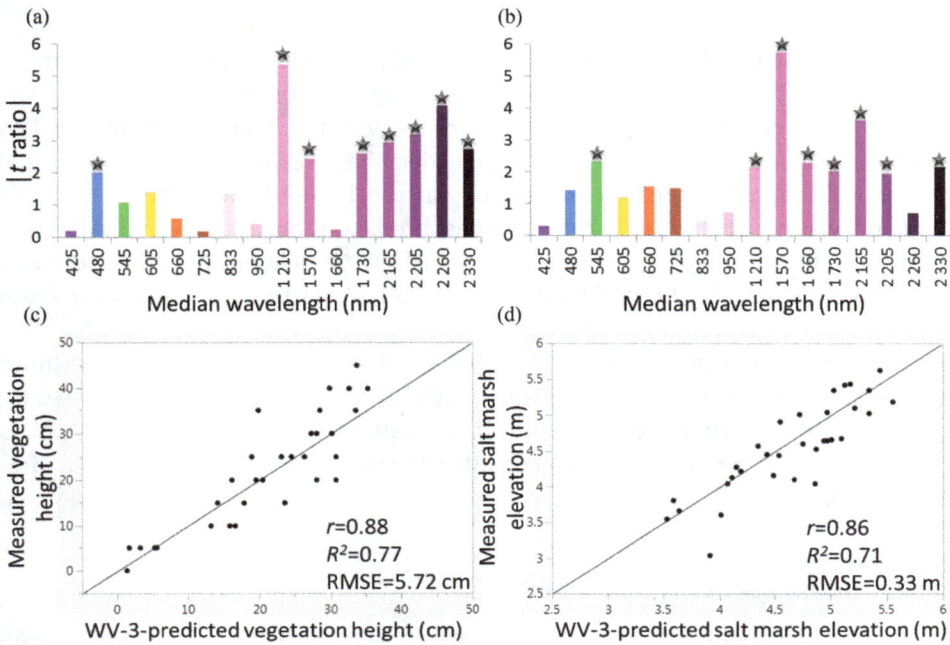

Figure 5. Bar plots of the absolute *t* ratios resulted from the prior linear regression of the 96 (a) vegetation height and (*b*) vegetation elevation samples, with respect to the 16 WorldView-3 spectral bands. Scatterplots of the 32 validation (c) vegetation height and (d) salt marsh elevation samples, as a function of the artificial neural network modelled (one hidden layer with seven neurons) points, derived from the best WorldView-3 spectral predictors (silver asterisks in a and b).

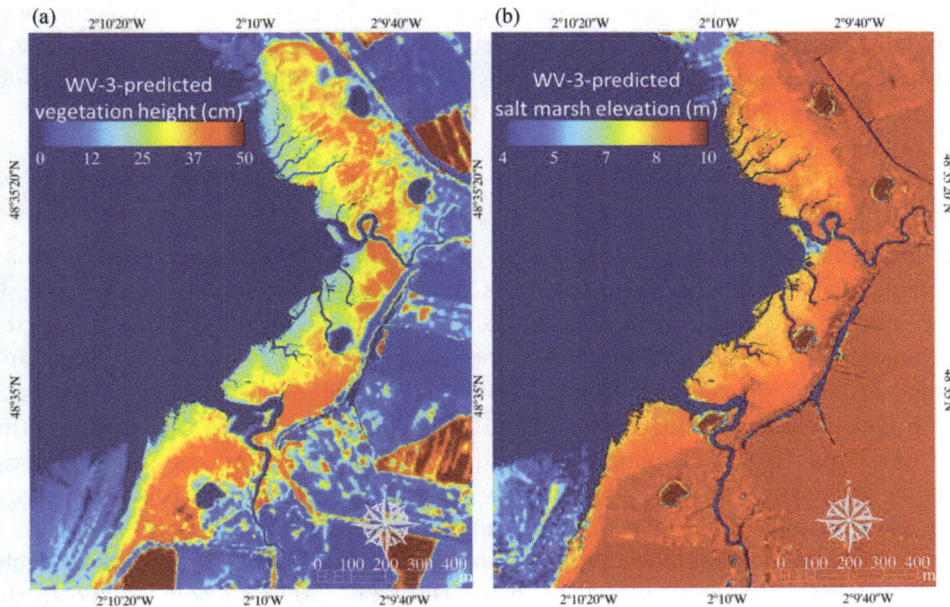

Figure 6. Digital (a) height and (b) elevation (MSL/IGN69) artificial neural network models (one hidden layer with seven neurons) derived from the best WorldView-3 spectral predictors (4318 × 5573 pixels at 0.31 m pixel size).

3.2. *Salt marsh elevation*

The analysis of the best WV-3 explanators of the salt marsh elevation revealed a stronger contribution of the SWIR gamut compared to the optical spectrum (Figure 5(b)). Only the SWIR-7 did not surpass the 0.05 significance value (t ratio = 0.7) in the SWIR predictors, whereas only the green band exceeded the threshold value (t ratio = 2.32) in the optical range. The greatest explanator was the SWIR-2 band (t ratio = 5.73). The ANN model was constrained with the eight best predictors, one hidden layer with seven neurons, and the elevation response composed of the 64 training samples. The shaped model, assessed with the 32 validation samples, had satisfactory performance (r = 0.86, $R2$ = 0.71, and RMSE = 0.33 m, Figure 5(d)).

The ANN model was implemented to each pixel of the WV-3 sub-data set, constituted of the green, SWIR-1, SWIR-2, SWIR-3, SWIR-4, SWIR-5, SWIR-6, and SWIR-8 bands so that the DEM is spatially explicit rendered at 0.31 m spatial resolution across the salt marsh (Figure 6(b)). The vegetation elevation consistently follows the salt marsh structure from lower to higher marsh.

3.3. *Salt marsh species composition*

The six supervised classifications joining the ANN models with some variations of the WV-3 reflectance and vegetation models were run to map the 15 classes (Figure 7). For the WV-3 sole bands (Figure 7(a-c)) and WV-3 bands enriched with previous models (Figure 7(d,f)), the increase in the number of spectral bands (from the four traditional to the 16 superspectral through the eight multispectral) augmented the classification accuracy. For both traditional (Figure 7(a)) and superspectral (Figure 7(c)) WV-3 spectral data sets, the addition of the two models (Figure 7(d,f), respectively) increased the accuracy, but did not significantly change the accuracy of the multispectral data set (Figure 7(b-e)). The best classification performance was therefore attained with the 16 superspectral WV-3 bands enriched with the two WV-3-derived models (Figure 7(f), 97.60%), whereas the lowest classification accuracy was attributed to the four traditional WV-3 sole bands (Figure 7(a), 69.27%).

4. Discussion

The salt marsh environment is challenging to be efficiently mapped given its inherent high spatial and spectral variabilities due to the low stratum location within steep micro- environments and the mixed reflectance resulting from a spectral combination of the vegetation, soil, and water features (Adam, Mutanga, and Rugege 2010). The straightforward pan-sharpening method to process 16 bands at 0.31 m was helpful to address the spatial complexity inherent to salt marshes. This step deserves to be examined by testing the hyper-sharpening methodology, recently developed for WV-3 imagery (Kwan et al. 2017). Salt marsh exhibits a low reflectance in the visible (400–700 nm) and the SWIR (1300–2500 nm) gamut as a consequence of the vegetation pigment and water content absorbance, respectively; however, it shows a high reflectance in the NIR (700–1300 nm) domain due to the combination of the vegetation leaf structure and salt marsh soil (Kumar et al. 2001; Rosso, Ustin, and Hastings 2005; Figure 8(a)).

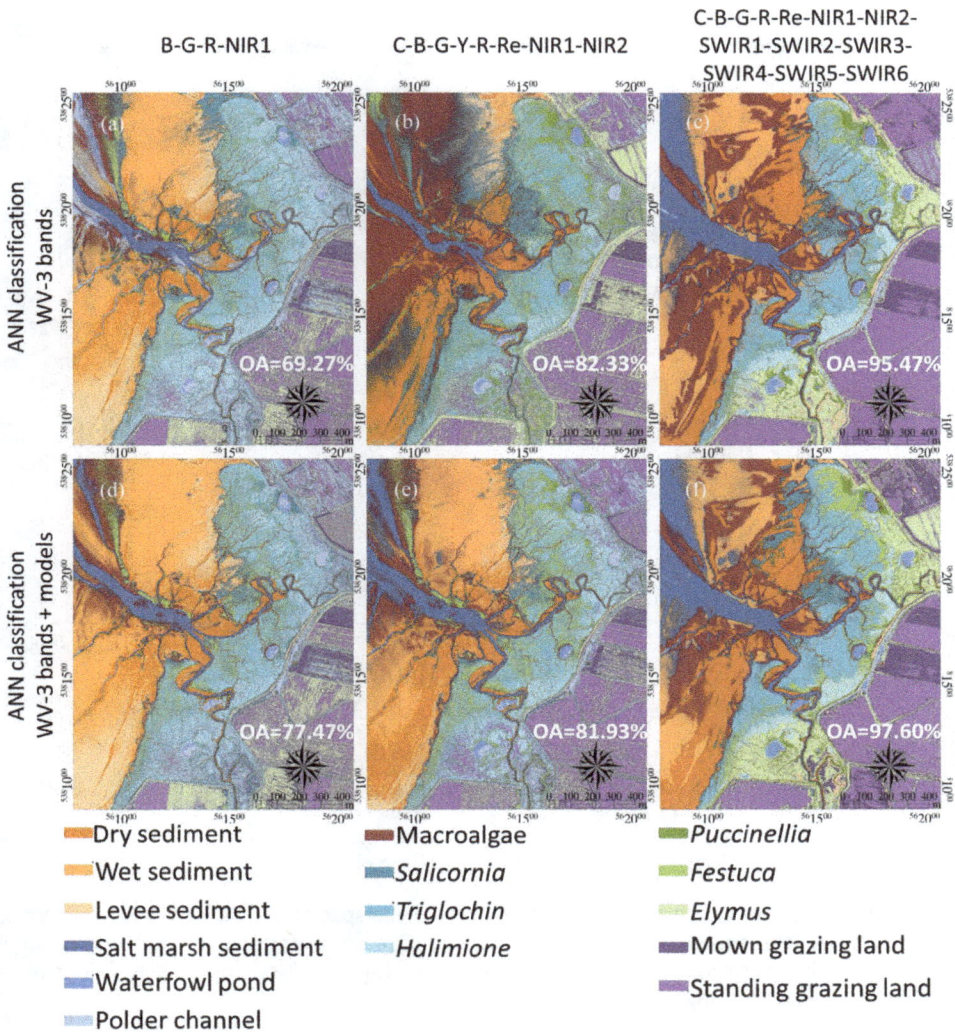

Figure 7. Classification of artificial neural network (ANN) models of the 15 classes based on 3000 training pixels (200 each class) and (a) four traditional, (b) eight multispectral, and (c) 16 super-spectral WorldView-3 (WV-3) spectral bands; and (d) four traditional WV-3 bands + digital vegetation height and elevation models (DVHM and DEM); (e) eight multispectral WV-3 bands + DVHM and DEM; and (f) 16 superspectral WV-3 bands + DVHM and DEM (4318 × 5573 pixels at 0.31 m pixel size). Overall accuracy (OA) of each ANN model was computed with 1500 validation pixels (100 each class). C, coastal; B, blue; G, green; Y, yellow; R, red; Re, red edge; NIRX, near-infrared X; SWIRX, shortwave infrared X.

4.1. *Salt marsh vegetation height*

Vegetation height component was first predicted by the NIR (SWIR-1), second by the SWIR (SWIR-2, and from SWIR-4 to SWIR-8), and finally by the blue bands (Figure 5(a)). The NIR, SWIR, and blue outcomes may be explained by the prevalence in reflectance of the chlorophyll vegetation over both marsh soil and marsh water signatures due to the

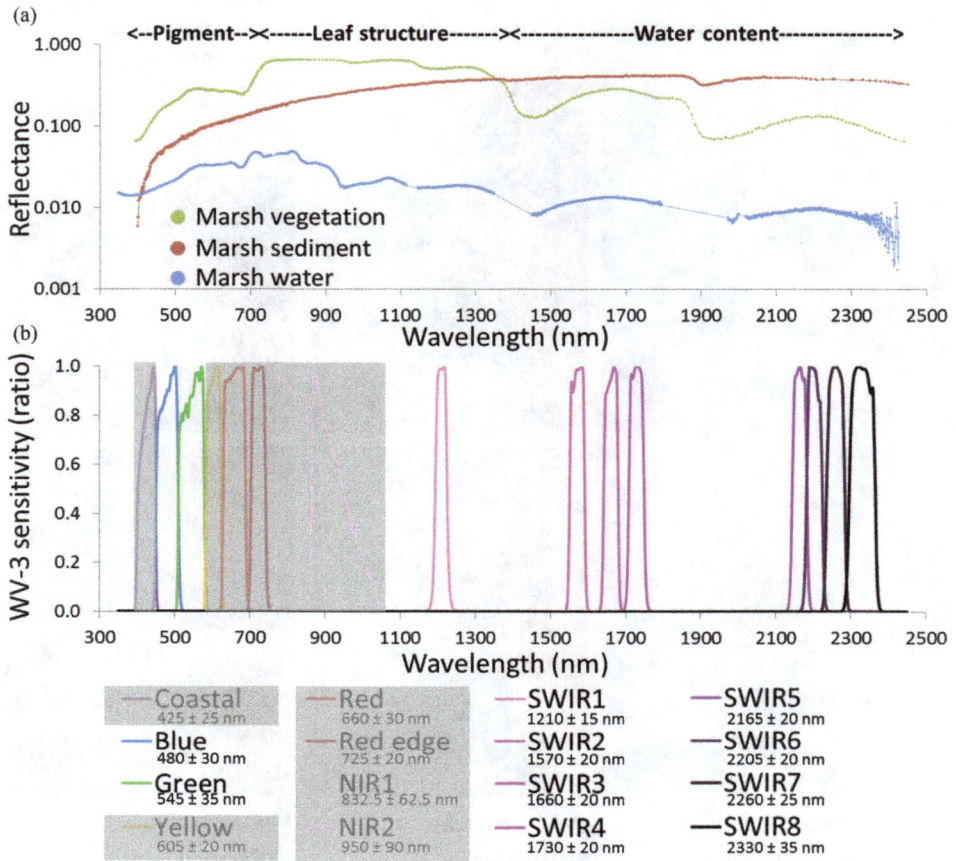

Figure 8. Line plots of (a) three salt marsh features' hyperspectral signatures (saltbrush *Atriplex garretti* marsh vegetation, grey silty clay marsh sediment, marsh water) derived from the handheld USGS spectral library version 7 (Kokaly et al. 2017), and (b) the 10 salt marsh-suited (out of 16) WorldView-3 spectral sensitivities (wavelength centre ± width/2 nm).

mesophyll (leaf) structure, the detection of the plant water absorption, and the peak of the chlorophyll absorption (Figure 8(a)). These assumptions rely on standard studies showing the covariance of the NIR with the plant moisture (1215 nm, Thenkabail et al. 2004); the SWIR with lignin, biomass, and water stress (Schmidt and Skidmore 2003, Thenkabail et al. 2004); and the blue with plant carotenoid concentration, as well as senescing and browning processes (495 nm, Thenkabail et al. 2004). The sole SWIR band that did not bring significant discrimination was SWIR-3 (1660 nm), and yet related to lignin and biomass (Thenkabail et al. 2004; Vaiphasa et al. 2005). Laboratory experiment will be likely to elucidate this issue, given that 1660 nm has been demonstrated to reveal a plant stress due to heavy metals (Thenkabail, Lyon, and Huete 2012). Further examination of the contribution of each highlighted WV-3 band per salt marsh class will establish valuable relationships between WV-3 spectral bands or combinations of bands and targeted salt marsh vegetation, soil, and water classes.

The salt marsh vegetation height model, based on the most informational WV-3 bands and *in situ*-calibrated ANN modelling, provided a low RMSE (i.e. 5.72 cm), which is better than another salt marsh vegetation height modelling method based on active lidar waveforms (i.e. 17.7 cm, Wang et al. 2009). That pioneer study pointed out a constant underestimation of the canopy height, which was not observed in the current ANN modelling process (Figure 5(c)). Such vertical accuracy spurs the combination of WV-3 and ANN to be transferred over external salt marshes but also any other terrestrial plant ecosystem composed of herbaceous strata. It is important to bear in mind that the WV-3 imagery requires to be ortho-rectified (for horizontal accuracy) with ground control points (GCPs), given a circular error at the 90th percentile (CE90) of 3.5 m.

4.2. *Salt marsh elevation*

Best WV-3 predictors used in marsh elevation ANN modelling were decreasingly found in the SWIR (from SWIR-2 to SWIR-6, and SWIR-8), the NIR (SWIR-1), and the green bands (Figure 5(b)). The spectral behaviour in the SWIR corresponds to a growing preponderance of the marsh soil signature over marsh plant and marsh water signatures (Figure 8(a)). The best predictors, peaking at 1570 and 2165 nm,and the longer SWIR bands might be related to signatures of plant litter (Thenkabail, Lyon, and Huete 2012), and lignin, biomass, and water stress (Schmidt and Skidmore 2003, Thenkabail et al. 2004). The value of the SWIR bands, sensitive to water, in elevation modelling suitably fits with the tidal flooding patterns (Pennings, Grant, and Bertness 2005): the higher the elevation is, the dryer the marsh is. The contribution of the green band, sensitive to chlorophyll (Thenkabail et al. 2004), might be explained by plant zonation, controlled by elevation, a distal driver of both emergence and salinity, wherever soil is not detected by the satellite. Only the SWIR-7 band (2260 nm) did not significantly improve the model in the SWIR range, which has yet been recognized as a spectral asset in detecting the soil background (Thenkabail, Lyon, and Huete 2012).

Coastal, yellow, red, red-edge, NIR-1, and NIR-2 bands (grey bands in Figure 8(b)) were found non-informational for both ANN models. The broadness of the NIR-1 and NIR-2 spectral resolutions (125 nm and 180 nm, respectively) might be proposed as a factor preventing them from accurately detecting the mesophyll structure (see Figure 8(a)). Atmospheric transmittance might be a control factor for the Coastal result. Further investigation is needed for explaining why salt marsh vegetation was not well modelled by yellow, red, and red-edge, usually valuable for crop vegetation detection (Thenkabail, Smith, and De Pauw 2002).

The salt marsh elevation model, based on the best WV-3 predictors and lidar-calibrated ANN, was validated with a moderate RMSE (0.33 m, Figure 5(d)), slightly higher than that of the nominal lidar error (0.2 m, Litto3D© 2015). This transferability potential is however impeded by other studies showing that lidar-derived salt marsh DEMs reached lower RMSEs (0.25 m, Hladik and Alber 2012) and even lower RMSEs when hyperspectral-derived classes were added for correction (from 0.15 to 0.1 m, Hladik, Schalles, and Alber 2013). Two other studies have found RMSE of lidar elevation, bottoming at 0.06 m (Morris et al. 2005, Wang et al. 2009).

4.3. *Salt marsh classification*

Extremely satisfactory vegetation height and satisfactory elevation models increased the OA classification of the 15 habitats, with the highest gain obtained for the four traditional WV-3 spectral data sets (+8.2%). As expected, the addition of four bands to the traditional data set strongly refined the classification of the eight multispectral data sets (+13.6% and +4.5% for data sets deprived of and provided with ancillary models, respectively). Likewise, the integration of eight SWIR bands to the multi-spectral data set considerably improved the classification of the 16 superspectral data sets (+13.1% and +15.7% for data sets deprived of/provided with ancillary models, respectively); thus, there was a growth of 26.2% and 20.1% from the four traditional to the 16 superspectral data sets (deprived of and provided with ancillary models, respectively). Whereas classifications stemming from the four traditional and eight multispectral data sets were lower than 82%, classifications based on the 16 superspectral data sets were greater than 95%, in agreement with the very high results computed for state-of-the-art normalized difference lidar vegetation index (NDLVI, 91.9% for 17 classes, Collin, Long, and Archambault 2010) and the fusion of hyperspectral and lidar elevation (90% for nine classes, Hladik, Schalles, and Alber 2013).

5. Conclusion

This innovative research has demonstrated that WV-3 imagery, when augmented at 16 bands (superspectral) and pansharpened at 0.31 m (VHR), is able to accurately map salt marsh elevation and vegetation height as well as species composition. Unprecedented findings can be summarized as follows.

(1) Salt marsh species composition (0.02 m spatial resolution) can be retrieved from air-truth (an airborne drone provided with a BGR camera).
(2) WV-3 visible, NIR, and SWIR reflectance bands can be pansharpened at VHR (0.31 m spatial resolution) with no significant degradation of radiometry.
(3) VHR WV-3 blue, SWIR-1, SWIR-2, SWIR-4, SWIR-5, SWIR-6, SWIR-7, and SWIR-8 bands are good predictors to model salt marsh vegetation height response using ANN ($r = 0.88$, $R2 = 0.77$, RMSE = 5.72 cm).
(4) VHR WV-3 green, SWIR-1, SWIR-2, SWIR-3, SWIR-4, SWIR-5, SWIR-6, and SWIR-8 bands are good predictors to model salt marsh elevation response using ANN ($r = 0.86$, $R2 = 0.71$, RMSE = 0.33 m).
(5) The VHR WV-3 superspectral (16 bands) data set better ANN-classified 15 salt marsh habitats than the multispectral (eight bands) data set, itself better than the traditional (four bands) data set (OA = 95.47, 82.33, and 69.27%, respectively).
(6) Adding WV-3-based salt marsh elevation and vegetation height models augmen-ted the ANN classification of the 15 classes for the superspectral and traditional data sets (OA = 97.60 and 77.47%, respectively), but not for the multispectral one (OA = 81.93%).

Acknowledgements

The authors gratefully thank DigitalGlobe Foundation for the courtesy of ortho-rectified Worldview-3 imagery. This work was supported by the French *Conservatoire* Du *Littoral et* Des *Rivages Lacustres*. Dorothée James and Hélène Gloria are also greatly acknowledged for their fieldwork involvement. The editor and two referees neatly improved the manuscript.

Disclosure statement

No potential conflict of interest was reported by the authors.

Funding

This work was supported by the Conservatoire du Littoral et des Rivages Lacustres [2016CV17].

References

Adam, E., O. Mutanga, and D. Rugege. 2010. Multispectral and Hyperspectral Remote Sensing for Identification and Mapping of Wetland Vegetation: A Review." *Wetlands Ecology and Management* 18 (3): 281–296. doi:10.1007/s11273-009-9169-z.

Allen, J. R. L. 2000. "Morphodynamics of Holocene Salt Marshes: A Review Sketch from the Atlantic and Southern North Sea Coasts of Europe." *Quaternary Science Reviews* 19 (12): 1155–1231. doi:10.1016/S0277-3791(99)00034-7.

Barbier, E. B., S. D. Hacker, C. Kennedy, E. W. Koch, A. C. Stier, and B. R. Silliman. 2011. "The Value of Estuarine and Coastal Ecosystem Services." *Ecological Monographs* 81 (2): 169–193. doi:10.1890/10-1510.1.

Belluco, E., M. Camuffo, S. Ferrari, L. Modenese, S. Silvestri, A. Marani, and M. Marani. 2006. "Mapping Salt-Marsh Vegetation by Multispectral and Hyperspectral Remote Sensing." *Remote Sensing of Environment* 105 (1): 54–67. doi:10.1016/j.rse.2006.06.006.

Bertness, M. D., P. J. Ewanchuk, and B. R. Silliman. 2002. "Anthropogenic Modification of New England Salt Marsh Landscapes." *Proceedings of the National Academy of Sciences* 99 (3): 1395–1398. doi:10.1073/pnas.022447299.

Casella, E., A. Collin, D. Harris, S. Ferse, S. Bejarano, V. Parravicini, V. J. L. Hench, and A. Rovere. 2017. "Mapping Coral Reefs Using Consumer-Grade Drones and Structure from Motion Photogrammetry Techniques." *Coral Reefs* 36 (1): 269–275. doi:10.1007/s00338-016-1522-0.

Chapman, M. G., and A. J. Underwood. 2011. "Evaluation of Ecological Engineering of "Armoured" Shorelines to Improve Their Value as Habitat." *Journal of Experimental Marine Biology and Ecology* 400 (1): 302–313. doi:10.1016/j.jembe.2011.02.025.

Collin, A., B. Long, and P. Archambault. 2010. "Salt-Marsh Characterization, Zonation Assessment and Mapping through a Dual-Wavelength LiDAR." *Remote Sensing of Environment* 114 (3): 520–530. doi:10.1016/j.rse.2009.10.011.

Collin, A., C. Ramambason, E. Casella, A. Rovere, V. Parravicini, L. Thiault, N. Nakamura, et al. 2018. "Mapping VHR Coral Reef Health Using Airborne Bathymetric LiDAR Elevation- Intensity Predictors, Visible Drone Response and Neural Network." *International Journal Remote Sensing Submitted.*

Collin, A., P. Archambault, and S. Planes. 2013. "Bridging Ridge-To-Reef Patches: Seamless Classification of the Coast Using Very High Resolution Satellite." *Remote Sensing* 5 (7): 3583–3610. doi:10.3390/rs5073583.

Collin, A., S. Etienne, and E. Feunteun. 2017. "VHR Coastal Bathymetry Using WorldView-3: Colour versus Learner." *Remote Sensing Letters* 8 (11): 1072–1081. doi:10.1080/2150704X.2017.1354261.

Congalton, R. G., and K. Green. 2009. *Assessing the Accuracy of Remotely Sensed Data: Principles and Practices*. Boca Raton, FL: CRC/Taylor Francis press.

Deegan, L. A., D. S. Johnson, R. S. Warren, B. J. Peterson, J. W. Fleeger, S. Fagherazzi, and W. M. Wollheim. 2012. "Coastal Eutrophication as a Driver of Salt Marsh Loss." *Nature* 490 (7420): 388. doi:10.1038/nature11533.

Gedan, K. B., B. R. Silliman, and M. D. Bertness. 2009. "Centuries of Human-Driven Change in Salt Marsh Ecosystems." *Annual Review of Marine Science* 1: 117–141. doi:10.1146/annurev. marine.010908.163930.

Heermann, P. D., and N. Khazenie. 1992. "Classification of Multispectral Remote Sensing Data Using a Back-Propagation Neural Network." *IEEE Transactions on Geoscience and Remote Sensing* 30 (1): 81–88. doi:10.1109/36.124218.

Hladik, C., J. Schalles, and M. Alber. 2013. "Salt Marsh Elevation and Habitat Mapping Using Hyperspectral and LIDAR Data." *Remote Sensing of Environment* 139: 318–330. doi:10.1016/j. rse.2013.08.003.

Hladik, C., and M. Alber. 2012. "Accuracy Assessment and Correction of a LIDAR-derived Salt Marsh Digital Elevation Model." *Remote Sensing of Environment* 121: 224–235. doi:10.1016/j. rse.2012.01.018.

Kokaly, R. F., R. N. Clark, G. A. Swayze, K. E. Livo, T. M. Hoefen, N. C. Pearson, R. A. Wise, et al. 2017. *USGS Spectral Library Version 7*. U.S. Geological Survey Data Series, Reston, VA, USA. doi:org/10. 3133/ds1035.

Kruse, F. A., and S. L. Perry. 2013. "Mineral Mapping Using Simulated Worldview-3 Short-Wave-Infrared Imagery." *Remote Sensing* 5 (6): 2688–2703. doi:10.3390/rs5062688.

Kumar, L., K. S. Schmidt, S. Dury, and A. K. Skidmore. 2001. "Review of Hyperspectral Remote Sensing and Vegetation Science." In *Imaging Spectrometry Basic Principles and Prospective Applications*, edited by F. D. Van Der Meer and S. M. De Jong. Dordrecht, The Netherlands: Kluwer.

Kwan, C., B. Budavari, A. Bovik, and G. Marchisio. 2017. "Blind Quality Assessment of Fused WorldView-3 Images by Using the Combinations of Pansharpening and Hypersharpening Paradigms." *IEEE Geoscience and Remote Sensing Letters* 14 (10): 1835–1839. doi:10.1109/ LGRS.2017.2737820.

Leconte, G. 1991. "Les Digues De Lancieux." In *Regards Sur Lancieux*, edited by J.P. Bihr, Saint-Jacut-de-la-Mer, Bretagne, France.

Litto3D© Version 1.0 – Content Description – March 2015(in French). retrieved on March 27, 2018, http://professionnels.ign.fr/doc/DC_Litto3D.pdf

Morris, J. T., D. Porter, M. Neet, P. A. Noble, L. Schmidt, L. A. Lapine, and J. R. Jensen. 2005. "Integrating LIDAR Elevation Data, Multi-Spectral Imagery and Neural Network Modelling for Marsh Characterization." *International Journal of Remote Sensing* 26 (23): 5221–5234. doi:10.1080/01431160500219018.

Mutanga, O., E. Adam, and M. A. Cho. 2012. "High Density Biomass Estimation for Wetland Vegetation Using WorldView-2 Imagery and Random Forest Regression Algorithm." *International Journal of Applied Earth Observation and Geoinformation* 18: 399–406. doi:10.1016/j.jag.2012.03.012.

Pennings, S. C., M. B. Grant, and M. D. Bertness. 2005. "Plant Zonation in Low-Latitude Salt Marshes: Disentangling the Roles of Flooding, Salinity and Competition." *Journal of Ecology* 93 (1): 159–167. doi:10.1111/j.1365-2745.2004.00959.x.

Rapinel, S., B. Clément, S. Magnanon, V. Sellin, and L. Hubert-Moy. 2014. "Identification and Mapping of Natural Vegetation on a Coastal Site Using a Worldview-2 Satellite Image." *Journal of Environmental Management* 144: 236–246. doi:10.1016/j.jenvman.2014.05.027.

Rosso, P. H., S. L. Ustin, and A. Hastings. 2005. "Mapping Marshland Vegetation of San Francisco Bay, California, Using Hyperspectral Data." *International Journal of Remote Sensing* 26: 5169–5191. doi:10.1080/01431160500218770.

Schmidt, K. S., and A. K. Skidmore. 2003. "Spectral Discrimination of Vegetation Types in a Coastal Wetland." *Remote Sensing of Environment* 85 (1): 92-108. doi:10.1016/S0034-4257(02)00196-7.

Thenkabail, P. S., E. A. Enclona, M. S. Ashton, and B. van Der Meer. 2004. "Accuracy Assessments of Hyperspectral Waveband Performance for Vegetation Analysis Applications." *Remote Sensing of Environment* 91: 354–376. doi:10.1016/j.rse.2004.03.013.

Thenkabail, P. S., J. G. Lyon, and A. Huete. 2012. *Hyperspectral Remote Sensing of Vegetation*. Boca Raton, FL, USA: CRC press book, Taylor and Francis Group.

Thenkabail, P. S., R. B. Smith, and E. de Pauw. 2002. "Evaluation of Narrowband and Broadband Vegetation Indices for Determining Optimal Hyperspectral Wavebands for Agricultural Crop Characterization." *Photogrammetric Engineering and Remote Sensing* 68: 607–621.

Vaiphasa, C., S. Ongsomwang, T. Vaiphasa, and A. K. Skidmore. 2005. "Tropical Mangrove Species Discrimination Using Hyperspectral Data: A Laboratory Study." *Estuarine, Coastal and Shelf Science* 65: 371–379. doi:10.1016/j.ecss.2005.06.014.

Wang, C., M. Menenti, M. P. Stoll, A. Feola, E. Belluco, and M. Marani. 2009. "Separation of Ground and Low Vegetation Signatures in LiDAR Measurements of Salt-Marsh Environments." *IEEE Transactions on Geoscience and Remote Sensing* 47 (7): 2014–2023. doi:10.1109/TGRS.2008.2010490.

Mapping semi-natural grassland communities using multi-temporal RapidEye remote sensing data

Christoph Raab ⓘ, H. G. Stroh, B. Tonn, M. Meißner, N. Rohwer, N. Balkenhol and J. Isselstein

ABSTRACT

Mapping semi-natural grassland has become increasingly important with regard to climate variability, invasive species, and the intensification of land use. At the same time, adequate field data collection is of pivotal importance for national and international reporting obligations, such as the European Habitats Directive. We present a remote-sensing-based monitoring framework for a Natura 2000 site with a heterogeneous composition of different grassland communities, using the Random Forest algorithm. Automated training data selection was successfully implemented based on the Random Forest proximity measure (Overall Accuracy ranging from 77.5–86.5%). RapidEye acquisitions originating from the onset of vegetation (prespring and first spring) and senescence (late summer and first autumn) were identified as important phenological phases for mapping semi-natural grassland communities. The derived probability maps of occurrences for each grassland class captured transitions between grassland communities and are therefore a better approximation of real-world conditions compared to classical, discrete maps.

1. Introduction

Semi-natural grasslands are habitats with high biodiversity (Dengler et al. 2014). They are characterised by indigenous, naturally occurring plant communities which have not been substantially modified, e.g. by sowing or fertilization. Unlike natural grassland, semi-natural grasslands have their origin in human activities, such as mowing or grazing, and depend on active management for their conservation (Peeters et al. 2014). Mapping and monitoring these habitats with their structural and botanical heterogeneity at very fine scales is a challenging task but becomes increasingly relevant with regard to the intensification of land use, land abandonment, climate variability, and invasive species (Stenzel et al. 2014; Wachendorf, Fricke, and Möckel 2017). National and international nature conservation and management activities, such as the European Habitats Directive

(Council Directive 92/43/EEC 1992) may even impose legal obligations to set up a monitoring framework for grasslands (Borre et al. 2011). Typically, grassland habitats are mapped and monitored through field surveys, which are time- and labour-intensive. In addition, they are difficult to reproduce, prone to subjective interpretation in the field, and in some cases limited by access restrictions. Remote-sensing-based mapping and monitoring offer unique possibilities to derive spatially explicit vegetation maps on large geographical areas using automated process chains (Borre et al. 2011; Buck et al. 2013; Nagendra et al. 2013; Stenzel et al. 2014; Corbane et al. 2015). Thus, remote-sensing-derived products can easily be updated at regular intervals.

The availability of remote sensing platforms applicable for land monitoring has experienced a tremendous boost since the 1970s (Belward and Skøien 2015). The increased amount of remote sensing images, however, requires effective supervised machine learning algorithms, such as Support Vector Machines (Cortes and Vapnik 1995; Mountrakis, Im, and Ogole 2011) or Random Forests (Breiman 2001; Belgiu and Drăguţ 2016) to extract relevant information from high-dimensional spectral data. The Random Forest classifier can be described as an ensemble of decision trees, from which the prediction is drawn by a majority vote. In contrast to Maximum Likelihood classifiers, no assumptions about the distribution of the data are required for the non-parametric Random Forest. Moreover, the Random Forest algorithm is insensitive to overfitting and its good performance for mapping vegetation has been demonstrated in several studies (Gislason, Benediktsson, and Sveinsson 2006; Cutler et al. 2007; Rodriguez-Galiano et al. 2012; Feilhauer et al. 2014; Barrett et al. 2016; Maxwell, Warner, and Fang 2018).

All supervised machine learning algorithms require a set of training data, which adequately represent the spectral characteristics of targeted vegetation classes. The training sampling strategy impacts classification accuracy and often leads to either over- or underrepresentation of classes in the final map (Millard and Richardson 2015; Ustuner, Sanli, and Abdikan 2016). Manual training sample selection is a time-consuming and subjective task (Rocchini et al. 2013), and so is the collection of test samples in the field, especially at transitions between different vegetation cover types.

Automated training data generation, derived, e.g. from existing reference maps, can be a potential solution to maintain objectivity in a cost- and labour-efficient way. For this, robust computer algorithms are needed to screen the initial training data set for incorrectly labelled samples, also called outliers. The Random Forest proximity of two training samples is a function of the ratio between the number of trees in which both samples share the same terminal node and the total number of trees in the forest (Belgiu and Drăguţ 2016). As pointed out by Gislason, Benediktsson, and Sveinsson (2006), Verikas, Gelzinis, and Bacauskiene (2011) and Touw et al. (2012), this measure can be used to detect and consequently exclude outliers in a training data set. Thus, by using the Random Forest proximity measure, uncertainties (e.g. at the transition between two vegetation communities) introduced by the training data sampling strategy or reference field mapping can be reduced.

In addition to the training data sampling design, the acquisition time of remote sensing data can be seen as an important factor influencing the quality of the mapping result (Nitze, Barrett, and Cawkwell 2015). Multi-seasonal remote sensing time series can support the discrimination of spectrally very similar land cover types, such as semi-natural grassland types, by incorporating temporal characteristics (Schmidt et al. 2014).

The identification of important temporal windows for classifying land cover types is not only of value to decrease model complexity and computation intensity, but also allows for a very targeted study design.

In this article, we explore the use of multi-annual, multi-seasonal remote sensing data combined with the Random Forest machine learning algorithm to discriminate semi-natural grassland communities identified by field mapping. The aims of this study were to:

- develop a framework to automatically derive training data by reducing uncertainties introduced by field mapping and sampling strategy,
- map the probability of spatial occurrence of different grassland communities,
- identify phenological seasons supporting the discrimination of different grassland communities.

The work was divided into a field mapping and data pre-processing part followed by the training data selection process. For each grassland class, the probability of occurrence was derived. Finally, permutation-based variable importance (Ruß and Brenning 2010) were calculated. The presented processing framework will help to improve current and future mapping and monitoring obligations, such as required by the European Habitats Directive.

2. Materials and methods

In this section, we provide an introduction to the study area and field mapping process, followed by our main study aim: automated selection of training data. In addition, we present how the derived training data can be applied for mapping the probability of occurrence of grassland at the community level. Important temporal windows for mapping semi-natural grassland are estimated using a permutation-based approach. A conceptual overview of methods applied in this manuscript is given in Figure 1.

2.1. *Study area*

The Grafenwoehr military training area (GTA) is located in the south-east of Germany (Bavaria) and lies at about 445 m above sea level in the natural region Upper Palatine–Upper Main Hills (Figure 2). Long-term annual averages of temperature and precipitation are $8.3 \pm 0.04°C$ and 701 ± 4 mm, respectively (1981–2010, mean ± SEM, of four weather stations of the German Weather Service (DWD, Deutscher Wetterdienst) in the immediate vicinity). The GTA extends about 223 km^2. Roughly 85% are part of the Natura 2000 network and contain numerous rare and highly protected habitat types, forming a refuge for many endangered species (Warren and Büttner 2008a, 2008b; Riesch et al. 2018). About 130 km^2 of the GTA are covered with forest and around 63 km^2 with semi-natural grasslands. The majority of the open areas are mown or mulched once per year between July and August. Fire and wildlife grazing, especially by red deer (*Cervus elaphus*), also play a role in some of these areas (Meißner et al. 2012).

In order to demonstrate the feasibility of monitoring semi-natural grassland communities via remote sensing in areas with limited access, we focused on two study sites,

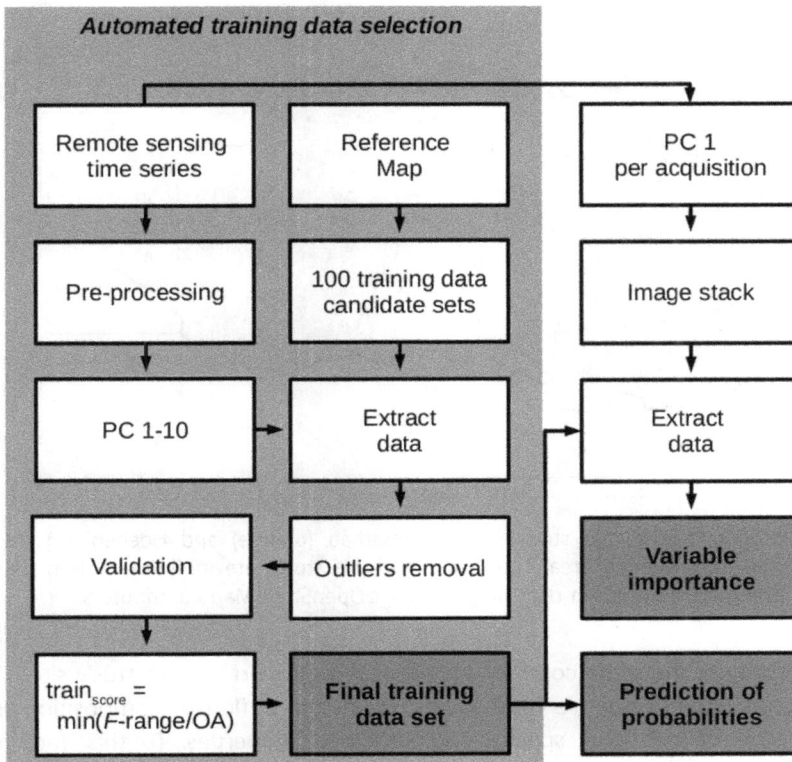

Figure 1. Schematic illustration of the automated training data selection process highlighted by a grey box. The right part of this figure emphasises the application of the derived training data set to calculate spatially explicit probability maps and to identify important phenological seasons for the discrimination of different grassland communities.

Sommerhau and Hoehenberg. Situated in the north of the GTA (Figure 2), about 71 ha of the total Hoehenberg site can be described as heterogeneous grassland, surrounded by forest. The northern two-thirds of the Hoehenberg site consist of a plain, underlain by Keuper sandstone, with only slight differences in the relief. Parts of the area are mulched once a year around the middle of July, mainly for fire safety reasons.

In contrast, the Sommerhau site is situated in the western part of the GTA (Figure 2). This area of about 140 ha is characterised by a mixture of grassland, fallow land and hedgerows, mainly underlain by more or less calcareous rock. In addition to some smaller forest stands, a larger continuous forested area can be found in the south. Compared to the Hoehenberg site, more anthropogenic infrastructure is present and the management regime includes mowing once a year around the middle of July.

2.2. Field mapping

Field data were collected at the two study sites between 2015 and 2017, mainly during summer. The collection of field data was limited by access restrictions due to military training activities. Even though geographical (Figure 2) and management differences exist,

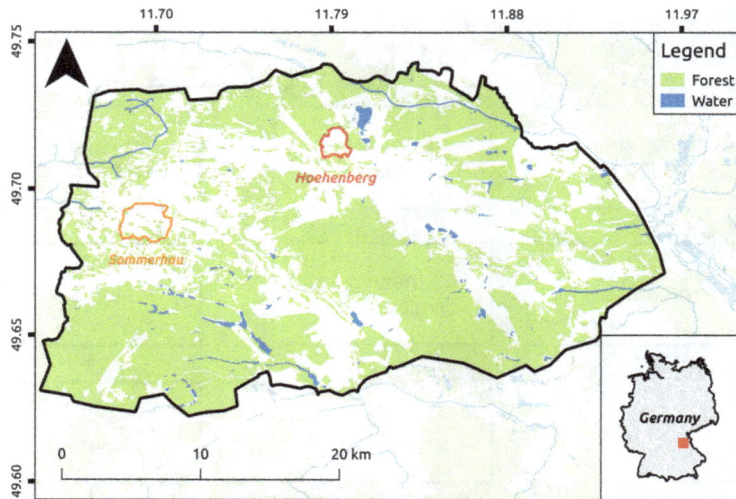

Figure 2. Location of the two study sites Sommerhau (orange) and Hoehenberg (red) in the Grafenwoehr military training area. The location of the study site in Germany is marked with a red square. The map is based on data provided by ©OpenStreetMap contributors.

a large quantity of the same common species can be found in both study sites. The field data collection aimed to derive vegetation units based on floristic composition as well as on dominance of individual species and structural properties. To this end, reference relevés were surveyed according to the EU Habitats Directive habitat types of Bavaria (BayLFU 2012). A selection of plant sociological literature aided the mapping process to adequately depict special local habitat types (Dierschke 1997; Dierschke and Briemle 2002; Burkart et al. 2004), as a restriction to the units in the Habitat Directive would have led to an inadequate depiction of unlisted local habitat types. Vegetation units (communities) were formed based on the surveyed reference plots and are outlined in Tables 1 and Tables 2, respectively. To derive a spatial grassland community map for each site, spatial boundaries were drawn according to the collected relevés data and visual interpretation in the field. This was supported by an aerial image taken during the midsummer season.

Table 1. Mapped grassland communities and share of total area for Sommerhau in percentage (%).

ID	Description	Share (%)
SG1	**Mown grassland dominated by short grasses (< 40%) – pasture-like lowland hay meadow**	**39.4**
SG11	Species-rich pasture-like lowland hay meadow, cover of *Trifolium repens* < 10%	16.0
SG12	Species-rich pasture-like lowland hay meadow, cover of *Trifolium repens* > 10%	22.0
SG13	Species-rich o pasture-like lowland hay meadow with species of dry calcerous grassland	0.5
SG14	*Lolium-perenne-Trifolium repens* pasture, cover of *Trifolium repens* > 10%	1.0
SG2	**Mown grassland dominated by tall grasses (> 40%) – typical lowland hay meadow**	**35.4**
SG21	Species-rich lowland hay meadow, cover of *Trifolium repens* < 10%	19.1
SG22	Species-rich lowland hay meadow with nutrient indicator species, cover of *Trifolium repens* > 10%	7.3
SG24	Species-rich or species-poor, disturbed lowland hay meadow	3.9
SG25	Species-rich lowland hay meadow with high cover of ruderal species	5.0
SB	**Fallow grassland**	**25.2**
SB10	Species-rich grassland fallow land, mesophilic sites	12.1
SB30	Species-poor grassland fallow land, mesophilic sites	11.3
SB40	Grassland fallow land, wet to dry locations	1.7
	Total area	138.4 ha

Table 2. Mapped grassland communities and share of total area for Hoehenberg in percentage (%).

ID	Description	Share (%)
HM1	**Vegetation dominated by short grasses, cover of tall grasses < 15%**	**47.1**
HM11	Species-rich *Festuca-rubra-Agrostis capillaris* meadow	13.9
HM12	Species-rich *Festuca-rubra-Agrostis capillaris* meadow, with grazing indicators	9.6
HM14	Species-rich *Festuca-rubra-Agrostis capillaris*, transition to *Nardus* grassland	20.9
HM15	Species-rich *Festuca-rubra-Agrostis capillaris* meadow, transition to oligotrophic siliceous grassland	2.8
HM2	**Vegetation dominated by tall grasses > 20%**	**27.5**
HM21	Species-rich meadow-like grassland	27.5
HM3	***Nardus* grassland**	**1.9**
HM30	*Nardus* grassland, cover of *Calluna vulgaris* < 30%	1.9
HM4	**Oligotrophic siliceous grassland**	**0.9**
HM40	Oligotrophic siliceous grassland	0.9
HB	**Fallow grassland**	**22.5**
HB50	*Calamagrostis-epigejos*-dominated	7.5
HB60	*Molinia* meadows, *Molinia caerulea* > 40%	8.1
HB70	Eutrophic wet grasslands, *Cyperaceae* > 50%	6.9
	Total area	71.0 ha

2.3. *Satellite data and pre-processing*

We acquired a multi-annual RapidEye time series (2014 – 2017) of 17 images (Table 3) with different temporal coverage for the Sommerhau and Hoehenberg sites. The processing level 3A that was used is already radiometrically, geometrically, and sensor corrected, and is delivered as 25 by 25 km tiles (ID-3,262,023).

Launched in 2008, the RapidEye satellite constellation consists of five identical satellites with a theoretically daily off-nadir recording interval (5.5 days at nadir) (Tyc et al. 2005). The high spatial resolution of 6.5 m pixel size is resampled to 5 m during the ortho-rectification by the data provider. In addition to the visible part of the electromagnetic spectrum, blue (440–510 nm), green (520–590 nm) and red (630–685 nm), the RapidEye satellites acquire top-of-atmosphere radiation in the rededge (690–730 nm) and near-infrared (NIR, 760–850 nm) part.

Table 3. Multi-annual RapidEye time series ordered by adjusted DOY (Julian day of the year). Study sites: Hoehenberg (H) and Sommerhau (S).

No	Acquisition date	Study site	Actual DOY	Adjusted DOY	Phenological phase
1	14 March 2016	H,S	74	88	Prespring (PSP), DOY 71–102
2	18 March 2015	S	77	89	
3	18 March 2016	H,S	78	91	
4	27 March 2017	H	86	95	
5	2 April 2014	H,S	92	114	First spring (FIS), DOY 103–131
6	20 April 2016	H,S	111	119	
7	17 April 2014	H,S	107	127	
8	22 Mai 2016	H,S	143	149	Full spring (FUS) DOY 132–158
9	26 Mai 2017	H	146	154	
10	11 June 2017	H,S	162	169	Early summer (ESU), DOY 159–179
11	8 June 2014	H,S	159	174	
12	24 June 2016	H,S	176	183	Midsummer (MSU), DOY 180–120
13	1 August 2016	H,S	214	221	Late summer (LSU), DOY 121–244
14	5 August 2015	H,S	217	227	
	–	–	–	–	Early autumn (EA), DOY 245–266
15	28 September 2014	S	271	277	Full autumn (FA), DOY 267–284
16	12 October 2015	H,S	285	287	Late autumn (LA), DOY 285–305
17	31 October 2015	H	304	300	

Temporal differences in phenological phases across years are a challenge when working with multi-annual time-series remote sensing data (Schmidt et al. 2014). Climate and local weather conditions affect the entry times of vegetation stages (phenological phases/seasons), therefore two satellite images from the same date but different years covering the study area may capture different phenological seasons. To account for this, Foerster et al. (2012) introduced a correction approach, which was successfully applied to multi-annual RapidEye data by Schmidt et al. (2014). To consider shifts in the phenology, we used observations of plant phenological seasons from the DWD, within a buffer distance of up to 30 km from the GTA centroid, to calculate the average entry time of each phenological season (Table 3) for the years 1951–2013. To estimate the deviation of each study year 2014–2017 from the long-term recordings, the averaged values were fitted to actual phenological observations of the respective years, using a third order polynomial (Schmidt et al. 2014). Based on these models, we calculated an adjusted Julian day of the year from the actual image acquisition dates (Table 3).

All images were reprojected to the German DHDN/3-degree Gauss–Kruger zone 4 (EPSG:31,468) reference system and co-registered (RMSE ≤ 0.5). Image correction included the transformation of raw Digital Numbers (DN) to radiance and top-of-atmosphere-reflectance (Planet Labs Inc 2016). Atmospheric correction was performed using the Second Simulation of Satellite Signal in the Solar Spectrum (6S) (Vermote et al. 1997) algorithm, as implemented in the *i.atcorr* function in the open source Geographic Resources Analysis Support System (GRASS GIS) version 7.2 (GRASS Development Team 2017). To account for distortions introduced by the topography, a C-factor correction (Teillet, Guindon, and Goodenough 1982) was applied using a sun illumination model derived from the European Digital Elevation Model (EU-DEM), version 1.1 (25 m spatial resolution).

2.4. *Training data sampling*

The training data selection was performed in a fully automated process chain, illustrated as a generalised flow chart in Figure 1. The key elements of the proposed procedure include a stratified random sampling based on a reference map (derived as described in section 2.2), the removal of potential outliers followed by validation and the selection of a final training data set.

As recommended by Colditz (2015), training points were randomly allocated proportional to the area covered by each class in the respective reference maps (Figure 3). Since we converted each pixel location into one point, no double sampling was possible. In addition, all non-grassland pixels were manually excluded from the study in advance by visual interpretation. To prevent a low representation of rare grassland communities, the minimum amount of training samples per class was set to 1% of the total number of training points (Colditz 2015). The total number of samples for each initial training data candidate set per study site was set to 15% of available pixels (8,453 pixels for Sommerhau and 4,261 for Hoehenberg). We repeated the random sampling 100 times, resulting in 100 initial training data candidate sets.

To reduce the complexity and to save calculation time, a Principal Component Analysis (PCA) was carried out for both study sites using all data available in the respective multi-seasonal RapidEye time series. Subsequently, the initial training data

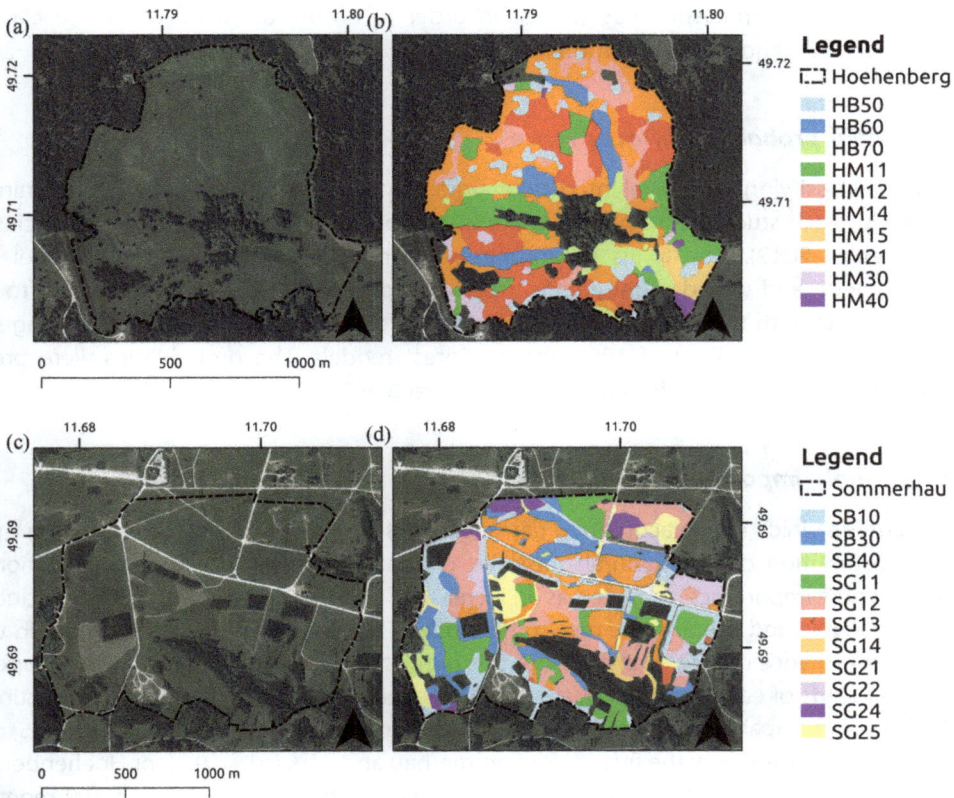

Figure 3. Aerial image (24 June 2016) of the Hoehenberg (a) and Sommerhau (c) (geodata based on: Bayerische Vermessungsverwaltung 2018). The result of field vegetation mapping for Hoehenberg (b) and Sommerhau (d). Explanations of the legend items can be found in Tables 1 and Tables 2.

candidate sets were used to extract the first ten Principal Component (PC) bands (PC1-10), explaining about 96% of the time series data for both Sommerhau and Hoehenberg.

Potential outliers were excluded based on the sample proximity measurement provided by the Random Forest (Breiman 2001; Gislason, Benediktsson, and Sveinsson 2006). The reduction of the 100 initial training data candidate sets was carried out using the function *rfOutliers* (threshold = 10) implemented in the *CORElearn package* (Robnik-Šikonja, Savicky, and Adeyanju Alao 2017) in R (R Core Team 2017).

To estimate the performance of each reduced training data candidate set we applied a 5-fold cross-validation approach with 100 repetitions using the *mlr* package (Bischl et al. 2016) and the Random Forest algorithm implemented in the *ranger* package (Wright and Ziegler 2015). In this study, all Random Forest models were constructed with 500 trees (num.trees) with the number of variables randomly sampled as candidates at each split (mtry) set to the square root of the number of input variables (Belgiu and Drǎguţ 2016).

The decision for the final training data set was based on the ratio of overall accuracy and the range of a class-specific performance measure (*F*-score). The *F*-score is calculated as the harmonic mean of user's and producer's accuracy, derived from the confusion matrix. The used range refers to the respective *F*-score values for all considered classes.

This train$_{score}$ ratio measure was chosen in order to ensure a balance between class-related accuracy and overall performance.

2.5. *Spatial probability of occurrence*

Instead of classifying each pixel into discrete grassland classes, we used the final training data set for each study site to predict the spatial probability of occurrence for each class (Malley et al. 2012). Due to its randomness, a constructed Random Forest model is always unique and would slightly differ from a second model that is constructed from the same data with the same settings. To account for these uncertainties, we averaged the predicted probabilities of 100 Random Forest models. The probabilities were predictions using the first ten PC bands of the respective RapidEye time series.

2.6. *Variable importance*

To determine which image acquisition dates (Table 3) are most relevant for estimating the spatial distribution of semi-natural grassland communities, we calculated permutation-based variable importance, recorded as an increase in classification error caused by excluding one variable and keeping the rest in the model. Thus, important dates contributing to model performance can be estimated. For this purpose, we compressed the main variance of all five bands of each image from the multi-seasonal RapidEye time series by extracting only the first Principal Component (PC1) per acquisition. On average, about 73% (sd = 6.5) of variance was explained by the first PC for Sommerhau and 84% (sd = 10.2) for Hoehenberg. Permutation was embedded in an additional cross-validation procedure with 100 repetitions, repeated 100 times on fold level, using the *mlr* and *ranger* packages.

3. Results

3.1. *Field mapping*

The field mapping revealed a high diversity of grassland communities for both study sites. Results of the field mapping are displayed in Figure 3. The mapped communities are outlined in Table 1 for Sommerhau and in Table 2 for Hoehenberg. The highest proportion of the Sommerhau site was covered with mown grassland dominated by short grasses (39.4%, SG11-SG14). The grassland class boundaries mostly followed the management (e.g. mowing). Fallow grassland classes could be found in the northern and western part, covering about one-quarter of the area (SB10-SB40). In contrast, a much more heterogeneous mosaic of less intensively managed grassland communities could be found in the Hoehenberg site. Short-grass dominated areas covered about one half of the site (HM11-HM15). With about 27%, stands dominated by tall grasses (HM21) were the most widespread class, occurring mainly in the northern two-thirds of the area.

3.2. *Training data sampling*

Using a proportional training data selection strategy along with a Random Forest proximity-based outliers detection, the mean number of points remaining in each of the 100 reduced training data candidate sets were 6,350 (sd = 134) points for Sommerhau and 3,331 (sd = 82) for Hoehenberg. On average 2,147 (sd = 133) outliers were excluded from the initial training data candidate sets for Sommerhau and 932 (sd = 83) for Hoehenberg. A detailed overview of class-specific values for both sites is outlined in Tables 4 and Tables 5, respectively. No outliers were detected for the classes SG12 and SG14, as well as for HM30 and HM40.

The average estimated performance of the 100 reduced training data candidate sets in terms of overall accuracy was about 9% higher for the Sommerhau (OA = 86%, Kappa = 0.84) compared to the Hoehenberg (OA = 78%, Kappa = 0.73) site (Table 6). The average F-score range was 8% higher for the Sommerhau compared to the Hoehenberg site.

The final training data set for all subsequent calculations was determined by the lowest F-score range:OA ratio. For Sommerhau the best training set in terms of this ratio resulted in an OA of 86.5% (sd = 0.2). A lower value was obtained for the Hoehenberg data set (OA: 77.5%, sd = 0.4). In comparison to the respective mean values of all training sets, the selected sample sets showed more balanced F-score values.

The cross-validated confusion matrix of the Random Forest classification applied on PC1-10 data of the RapidEye time series yielded good class-specific results for Sommerhau (Table 7) and Hoehenberg (Table 8). The lowest user's and producer's accuracies were estimated for relatively rare classes, such as HM40 and SG14.

Table 4. Results of the training data sampling procedure for all classes (Table 1) of the Sommerhau site, reported as the amount of points per class. The performance is given by the harmonic mean of user's and producer's accuracy (F-score). sd = standard deviation.

	SB10	SB30	SB40	SG11	SG12	SG13	SG14	SG21	SG22	SG24	SG25
Mean number of points	647	690	133	1049	1449	84	84	1164	445	231	374
sd	21	16	3	18	20	-	-	21	12	9	14
Mean number of outliers	357	260	8	279	373	-	-	418	164	96	192
sd	21	16	3	18	20	-	-	21	11	9	14
Mean F-score	82.2	80.6	68.3	92.4	87.4	77.6	66.0	88.3	81.3	86.8	86.1
sd F-score	1.3	1.0	3.7	0.6	0.6	2.9	4.6	0.7	1.6	1.6	1.2
Number of points of the selected training set	663	688	134	979	1400	84	84	1171	443	203	356
Number of detected outliers points of the selected training set	341	256	6	306	422	-	-	411	166	108	209
F-score of the selected training set	84.9	82.2	73.1	92.5	87.3	77.7	73.6	89.5	77.6	86.4	86.2
sd F-score of the selected training set	0.6	0.5	2.1	0.3	0.3	2.0	2.7	0.3	1.0	1.0	0.8

Table 5. Results of the training data sampling procedure for all classes (Table 2) of the Hoehenberg site, reported as the amount of points per class. The performance is given by the harmonic mean of user's and producer's accuracy (F-score). sd = standard deviation.

	HB50	HB60	HB70	HM11	HM12	HM14	HM15	HM21	HM30	HM40
Mean number of points	215	280	256	412	353	676	113	901	82	43
sd	10	7	5	15	8	16	2	19	-	-
Mean number of outliers	104	65	39	179	55	214	4	272	-	-
sd	10	7	5	15	8	16	3	19	-	-
Mean F-score	78.3	86.8	85.2	78.9	68.6	76.5	83.4	77	68.7	79.4
sd F-score	2.8	1.7	1.8	2.0	2.2	1.5	2.6	1.0	4.4	4.0
Number of points of the selected training set	206	286	251	431	349	671	115	916	82	43
Number of detected outliers points of the selected training set	113	59	44	157	59	219	2	255	-	-
F-score of the selected training set	77.7	84.9	83.3	79.2	73.6	72.7	83.5	76.8	74.5	81.7
sd F-score of the selected training set	1.0	0.7	0.9	0.9	1.0	0.8	1.2	0.4	2.4	2.7

Table 6. Results of 5-fold cross-validation with 100 repetitions of the 100 reduced training data candidate sets and the selected training data set. OA = overall accuracy, sd = standard deviation, F-range = range of class-specific F-scores.

	OA	OA sd	Kappa	Kappa sd	F-range	F-range sd	F-range/OA
Sommerhau selected training set	86.5	0.2	0.84	0.0027	20.1	2.1	0.23
Sommerhau mean of all models	86	0.4	0.84	0.005	27.9	6.7	0.32
Hoehenberg selected training set	77.5	0.4	0.73	0.0045	12.2	1.4	0.16
Hoehenberg mean of all models	78	0.7	0.73	0.0088	20.1	3.2	0.26

In addition, we analysed the spatial characteristics of the excluded outliers summed up for all initial training data candidate sets, illustrated as a density raster in Figure 4. For both sites, core areas with no outliers can be observed. The density of outliers for the Sommerhau site followed the spatial patterns of infrastructure, e.g. roads. Both cases provide evidence for a general edge effect, characterised by high density of outliers present at transitions between different grassland communities.

3.3. *Spatial probability of occurrence*

For each class, we derived the modelled probability of occurrence for each pixel, averaged over 100 repetitions. The exemplary results for the classes SG21 and SG14 for the Sommerhau site and HM21 and HB70 for the Hoehenberg site are illustrated in Figure 5. For all other grassland communities, the results are displayed in Figure S1-S17 in the supplementary material. Very high class probabilities can be observed within the field-mapped boundaries for all grassland communities and both sites. In general, areas with lower probabilities can be found mostly in the direct vicinity of the mapped boundaries, with a gradual decrease with distance.

3.4. *Variable importance*

Permutation-based variable importance, derived from an increase in classification error, were used to identify important acquisition dates. As illustrated in Figure 6, the importance across and within phenological seasons varied, with a maximum increase in classification error of about 20% for the first autumn (FA) image number 15 in Sommerhau. The cumulative importance for all prespring (PSP) acquisitions was much higher for Sommerhau than in Hoehenberg, but relatively similar between the two sites for the first spring (FIS) season. In general, for both sites, the late summer (LSU) season appeared to be relevant for mapping grassland at the community level, and for Sommerhau especially also the first autumn (FA). Unfortunately, no image for this period was available for a comparison with Hoehenberg.

In summary, the least important phenological seasons for mapping semi-grassland at the community level, in terms of an increase in classification error, were the full spring (FUS) and early summer (ESU) for both study sites.

4. Discussion

The field mapping of both study sites was challenged by access limitations and, given the heterogeneous grassland community composition, a labour-intensive and time-consuming task. Similar circumstances for field data collection are reported by Ghimire et al. (2012), and

Table 7. Confusion matrix of grassland communities classified by Random Forest applied on the first ten Principal Component bands of the Sommerhau RapidEye time series. Data is averaged over 100 cross-validation repetitions.

		Reference											Total	User's accuracy
		SB10	SB30	SB40	SG11	SG12	SG13	SG14	SG21	SG22	SG24	SG25		
Prediction	SB10	564	62	14	1	0	6	0	11	0	2	4	**664**	0.85
	SB30	74	596	38	0	0	1	0	41	1	1	3	**755**	0.79
	SB40	1	2	82	0	0	0	1	0	0	0	0	**86**	0.94
	SG11	12	7	0	943	33	1	6	5	3	5	2	**1017**	0.93
	SG12	0	1	0	68	1273	8	22	26	63	21	35	**1517**	0.84
	SG13	0	0	0	0	1	56	1	0	0	0	2	**60**	0.93
	SG14	0	0	0	0	3	0	51	0	0	0	0	**54**	0.94
	SG21	11	23	0	3	31	4	0	1068	61	10	5	**1216**	0.88
	SG22	0	0	1	0	29	0	0	19	312	0	3	**363**	0.86
	SG24	0	0	1	1	10	0	0	1	0	177	1	**191**	0.93
	SG25	0	2	0	5	19	7	2	1	2	3	302	**343**	0.88
	Total	**662**	**693**	**135**	**1021**	**1399**	**83**	**83**	**1172**	**442**	**219**	**357**		
	Producer's accuracy	0.85	0.86	0.61	0.92	0.91	0.67	0.60	0.91	0.71	0.81	0.85		

Table 8. Confusion matrix of grassland communities classified by Random Forest applied on the first ten Principal Component bands of the Hoehenberg RapidEye time series. Data is averaged over 100 cross-validation repetitions.

		Reference											
		HB50	HB60	HB70	HM11	HM12	HM14	HM15	HM21	HM30	HM40	Total	User's accuracy
Prediction	HB50	145	0	4	1	0	14	0	2	1	0	**167**	0.86
	HB60	4	234	9	1	1	3	0	12	0	0	**264**	0.88
	HB70	4	12	201	2	0	0	0	10	0	0	**229**	0.87
	HM11	2	1	1	320	1	24	6	12	6	2	**375**	0.85
	HM12	5	10	0	2	250	28	1	33	1	0	**330**	0.76
	HM14	10	4	1	45	29	473	6	55	9	0	**632**	0.75
	HM15	3	0	0	2	0	2	92	0	1	4	**104**	0.88
	HM21	31	25	34	56	66	126	6	794	9	0	**1147**	0.69
	HM30	1	0	0	2	2	1	0	0	53	1	**60**	0.88
	HM40	0	0	0	2	0	0	3	0	2	35	**42**	0.83
	Total	**205**	**286**	**250**	**433**	**349**	**671**	**114**	**918**	**82**	**42**		
	Producer's accuracy	0.70	0.82	0.80	0.74	0.72	0.71	0.80	0.86	0.65	0.81		

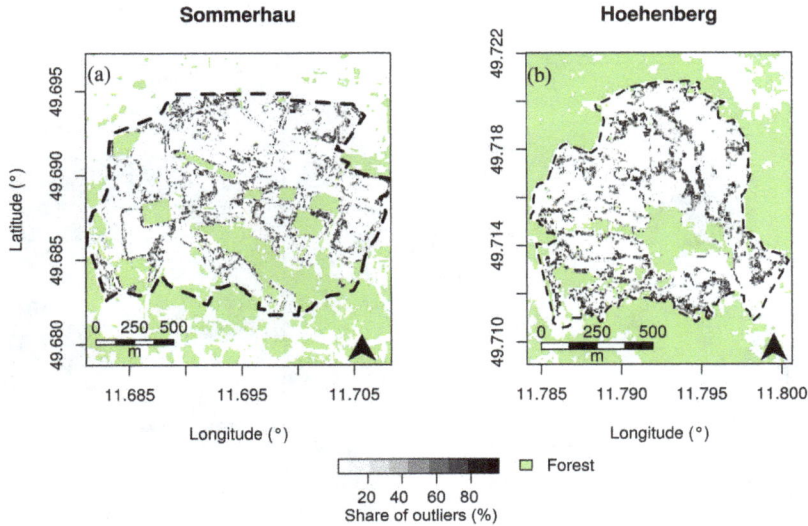

Figure 4. Share of outliers divided by the total amount of members of training data candidate sets per pixel expressed as percentage. Sommerhau (a) and Hoehenberg (b).

are probably quite common on many military training areas, which often support high levels of biodiversity and are of major management and conservation concern (e.g. Benton, Ripley, and Powledge 2008). Uncertainty in training data due to the subjectivity of field surveys, as well as the mixed pixel problem, are major challenges in remote sensing (Rocchini et al. 2013). In our study, the applied screening of training data for potential outliers excluded about 25% of the initial field reference set for both study sites. As the training data set were screened for outliers, potential uncertainties were reduced. High accuracies were estimated for the respective final training data set, based on the PC1-10 response variables (Sommerhau: OA = 86.5%, Kappa = 0.84 and Hoehenberg: OA = 77.5%, Kappa = 0.73). The performance with a Random Forest proximity threshold of 10 was good. The proximity measure can be related to the spectral distance of two samples and a more conservative, i.e. smaller, threshold value would lead to more homogeneous training points for the respective classes. For very low threshold values, the model would be trained with pure pixels only, and would therefore not be able to reliable classify mixed pixels. In general, there is a need to explore the sensitivity of results to the threshold setting for the Random-Forest-based outliers detection method (Belgiu and Drăguţ 2016). In addition, its potential to identify subclasses by grouping samples with similar proximities to all other training points of the same class should be explored in future research activities (Touw et al. 2012).

The spatial distribution of detected outliers for several iterations (Figure 4) highlights areas with potential mapping errors, as well as uncertainties about the discrete spatial boundaries between grassland communities drawn during the mapping process. These areas can be specifically addressed by the following mapping iteration to revise the spatial class boundaries. This can be especially useful for long-term monitoring strategies in combination with probability maps to detect potential early warning signals for the decline of a high-value grassland community in terms of conservation.

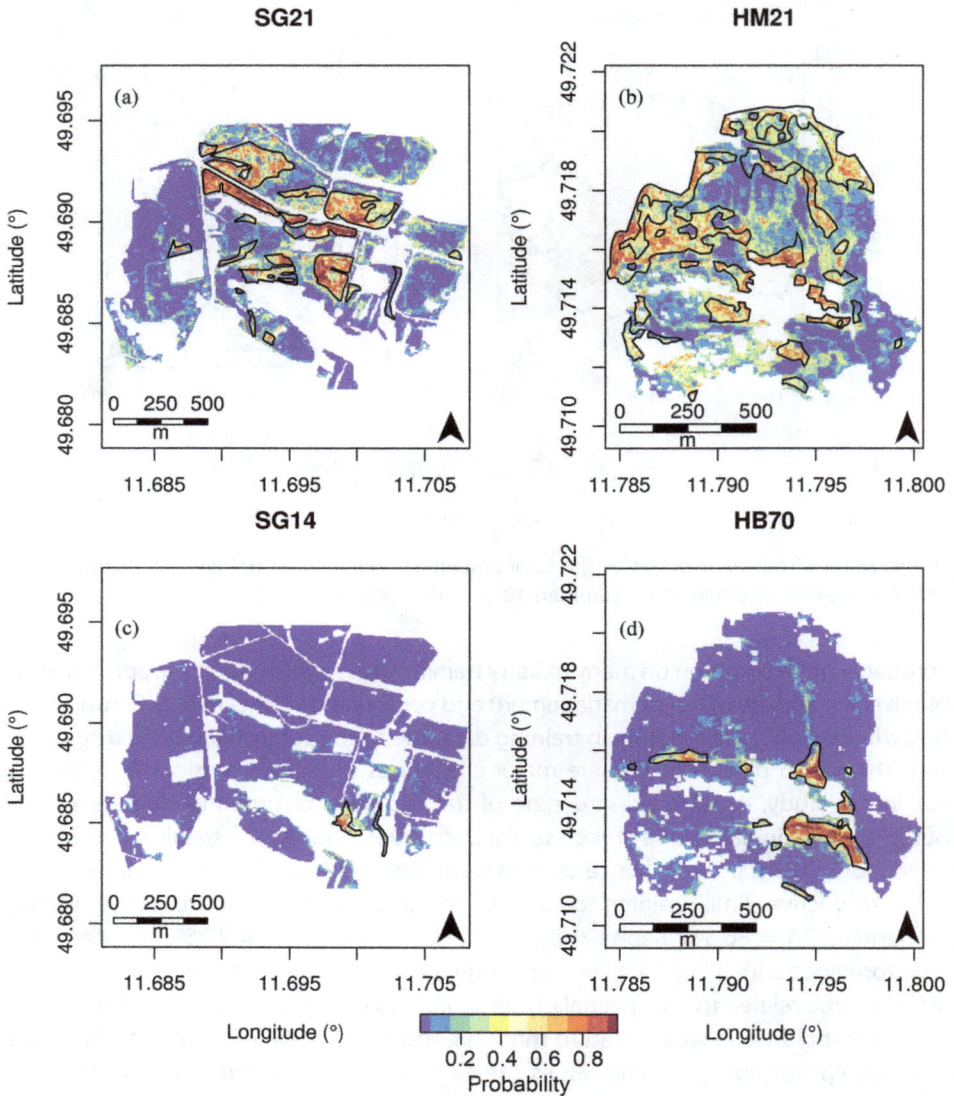

Figure 5. Exemplary results for the spatial probability of occurrence for the Sommerhau classes SG21 (a) and SG14 (c), as well as the Hoehenberg classes HM21 (b) and HB70 (d). The field mapping results are drawn as black polygons. Detailed descriptions of the classes can be found in Tables 1 and Tables 2.

The estimated performance during the training data selection procedure was consistently better for the Sommerhau site compared to Hoehenberg. This can be attributed to a more heterogeneous land cover for the Hoehenberg site and different land management practices. The amount of initial training points per study site (15% of the area) can be regarded as relatively high compared to common recommendations (Colditz 2015; Belgiu and Drăguţ 2016), even though about one quarter was subsequently excluded by the outliers screening. However, in contrast to a fixed number of

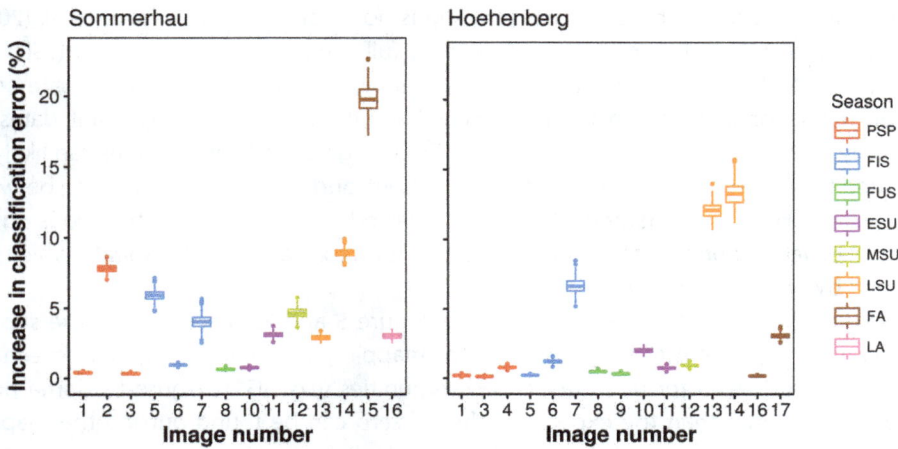

Figure 6. Permutation-based variable importance in terms of increase in classification error caused by excluding the PC1 band for one date and keeping the rest in the model. Explanations to the image number of the x-axis and the phenological seasons of the legend can be found in Table 3.

training points related to the study site size, Millard and Richardson (2015) recommended that the training set should be as large as possible and randomly distributed, while accounting for the proportion of land covered by each class. In addition, studies showed that the Random Forest benefits to a great extent from large sample sizes (Fassnacht et al. 2014; Ma et al. 2017).

The selection of the most important variables contributing to a classification model has gained a lot of attention over the last decade (Diaz-Uriarte 2007; Pal and Foody 2010; Verikas, Gelzinis, and Bacauskiene 2011; Rodriguez-Galiano et al. 2012; Chutia et al. 2017), and methods to select the most influential variables include backward elimination (Dash and Liu 1997) and forward selection strategies (Langley 1994). The Random Forest algorithm provides a relative measure of variable importance in the form of the change of an internal classification error measure and the Gini index caused by excluding one variable and keeping the rest in the model (Breiman 2001; Rodriguez-Galiano et al. 2012). These variable importance measures can, however, display a bias when variables are correlated (Strobl et al. 2008). In this case high importance can be observed in favour of insignificant variables correlated with significant variables at the cost of reduced importance measures for significant, uncorrelated variables (Nicodemus and Malley 2009; Conn et al. 2015). In addition, the Random Forest variable importance is not reliable when the predictor variables vary in scale or number of categories (Strobl et al. 2007). Introduced by Strobl et al. (2008), the conditional variable importance for Random Forest is one approach to address the discussed importance measure problems, but is known to be very calculation intensive (Nicodemus et al. 2010). Therefore, a permutation-based variable importance estimation can be seen as a valid alternative.

The results of the permutation-based variable importance of this study indicate a general importance of the first spring (FIS) and late summer (LSU) season for both sites. Particularly for Sommerhau, the first autumn (FA) can be seen as the most influential temporal window, with an estimated mean increase of about 20% in classification error when excluded from the model. Full spring (FUS) and early summer (ESU) were the least important phenological seasons for both sites in this study. This finding is not supported by Schmidt et al. (2014), who identified early summer (ESU) – followed by full spring (FUS), late summer (LSU) and midsummer (MSU) – as the most important phenological seasons to discriminate semi-natural grasslands in Northern Germany. However, the inconsistent important dates for accurately mapping northern versus southern German grassland communities can likely be attributed to differences in the location, management, and grassland composition between studied areas. However, Förster et al. (2010) suggested using RapidEye acquisitions during the onset of vegetation and senescence phases to monitor Natura 2000 habitats, which was indicated by our findings as well.

When the probability values of occurrence (Figure 5 and Figure S1-S17 in the supplementary material) are plotted together with the mapping results, a good spatial agreement can be observed, even for rare grassland communities (e.g. SG14, Figure 5). Some probability values higher than the expected value of zero can be found outside the mapped boundaries. For the example of class HB70, this can be linked to the topography, e.g. related to a ditch going from the north to the south. Areas with lower probabilities can be attributed to continuous transitions due to self-organisation in vegetation and changes in environmental conditions (Rocchini et al. 2013). While a common pixel-based classification of such gradual transitions into a fixed number of discrete classes would not have reflected the true reality of the Earth's surface, the proposed mapping approach captures gradual transitions by probabilities of occurrence. A segmentation-based mapping approach would most likely increase the overall accuracies for high spatial resolution data by aggregating spectrally homogeneous pixels into objects (Laliberte, Fredrickson, and Rango 2007; Förster et al. 2010), but with the trade-off of losing fine-scale information as small features are swallowed into bigger objects (Schmidtlein and Sassin 2004; Schmidt et al. 2017). In addition, it is easy to aggregate pixels to objects for reporting purposes, while the reverse is impossible.

5. Conclusion

In this study, we showed that training data for mapping and monitoring vegetation cover can automatically be derived in an objective way and that multi-annual, multi-seasonal remote sensing data can be successfully applied to monitor semi-natural grassland vegetation types at a fine scale. The introduced training data sampling framework can help to identify potential uncertainties in the reference data. Pivotal for this is the collection of baseline data by field mapping. With regard to the reporting obligations under Art.-17 of the EU Habitats Directive, the proposed mapping strategy can locate hot-spot areas of change by incorporating future remote sensing data. Thus, an effective fieldwork strategy can be designed to target areas of special interest. Full spring (FUS) and early summer (ESU) were identified as the least important phenological seasons for mapping semi-natural grassland. Future research should consider the synergistic possibilities of combining multi-spectral and radar data for monitoring semi-natural grassland (Schuster et al. 2011; Metz et al. 2012; Bargiel 2013).

Acknowledgments

The project was supported by funds of German government's Special Purpose Fund held at Landwirtschaftliche Rentenbank (28 RZ 7007). We thank the Federal Forests Division

(Bundesforst) of the German Institute for Federal Real Estate (Bundesanstalt für Immobilienaufgaben) and the Institut für Wildbiologie Göttingen und Dresden e.V. for close cooperation and support. We acknowledge the DLR for the delivery of RapidEye images as part of the RapidEye Science Archive (RESA) – proposal 00226.

Disclosure statement

No potential conflict of interest was reported by the authors.

Funding

This work was supported by the German government's Special Purpose Fund held at Landwirtschaftliche Rentenbank [28 RZ 7007].

ORCID

Christoph Raab ⓘ http://orcid.org/0000-0002-3044-6826

References

Bargiel, D. 2013. "Capabilities of High Resolution Satellite Radar for the Detection of Semi-Natural Habitat Structures and Grasslands in Agricultural Landscapes." *Ecological Informatics* 13: 9–16. doi:10.1016/j.ecoinf.2012.10.004.

Barrett, B., C. Raab, F. Cawkwell, and S. Green. 2016. "Upland Vegetation Mapping Using Random Forests with Optical and Radar Satellite Data." *Remote Sensing in Ecology and Conservation* 2 (4): 212–231. doi:10.1002/rse2.32.

BayLFU, (Bayerisches Landesamt für Umwelt). 2012. "Biotopkartierung - Kartieranleitungen - LfU Bayern." Kartieranleitungen. Accessed 6 October 2017. https://www.lfu.bayern.de/natur/biotop kartierung_flachland/kartieranleitungen/index.html.

Belgiu, M., and L. Drăguţ. 2016. "Random Forest in Remote Sensing: A Review of Applications and Future Directions." *ISPRS Journal of Photogrammetry and Remote Sensing* 114: 24–31. doi:10.1016/j.isprsjprs.2016.01.011.

Belward, A. S., and J. O. Skøien. 2015. "Who Launched What, When and Why; Trends in Global Land-Cover Observation Capacity from Civilian Earth Observation Satellites." *ISPRS Journal of Photogrammetry and Remote Sensing* 103: 115–128. doi:10.1016/j.isprsjprs.2014.03.009.

Benton, N., J. D. Ripley, and F. Powledge. 2008. "Conserving Biodiversity on Military Lands: A Guide for Natural Resources Managers." 2008 edition. In *Arlington (VA): NatureServe.* Available at http://www.dodbiodiversity.org.

Bischl, B., M. Lang, L. Kotthoff, J. Schiffner, J. Richter, E. Studerus, G. Casalicchio, and M. Z. Jones. 2016. "Mlr: Machine Learning in R." *Journal of Machine Learning Research* 17 (170): 1–5.

Borre, J. V., D. Paelinckx, C. A. Mücher, L. Kooistra, B. Haest, G. De Blust, and A. M. Schmidt. 2011. "Integrating Remote Sensing in Natura 2000 Habitat Monitoring: Prospects on the Way Forward." *Journal for Nature Conservation* 19 (2): 116–125. doi:10.1016/j.jnc.2010.07.003.

Breiman, L. 2001. "Random Forests." *Machine Learning* 45 (1): 5–32. doi:10.1023/A:1010933404324.

Buck, O., A. Klink, V. E. G. Millán, K. Pakzad, and A. Müterthies. 2013. "Image Analysis Methods to Monitor Natura 2000 Habitats at Regional Scales–The MS. MONINA State Service Example in Schleswig-Holstein, Germany." *Photogrammetrie-Fernerkundung-Geoinformation* 2013 (5): 415–426. doi:10.1127/1432-8364/2013/0188.

Burkart, M., H. Dierschke, N. Holzel, B. Nowak, and T. Fartmann. 2004. "Molinio-Arrhenatheretea (E1)-Teil 2: Molinietalia." *Synopsis* 9: 1–103.

Chutia, D., D. K. Bhattacharyya, J. Sarma, and P. N. L. Raju. 2017. "An Effective Ensemble Classification Framework Using Random Forests and a Correlation Based Feature Selection Technique." *Transactions in GIS* 21.6 (2017): 1165–1178. https://doi.org/10.1111/tgis.12268

Colditz, R. R. 2015. "An Evaluation of Different Training Sample Allocation Schemes for Discrete and Continuous Land Cover Classification Using Decision Tree-Based Algorithms." *Remote Sensing* 7 (8): 9655–9681. doi:10.3390/rs70809655.

Conn, D., T. Ngun, G. Li, and C. Ramirez. 2015. *Fuzzy Forests: Extending Random Forests for Correlated, High-Dimensional Data.* UCLA: Biostatistics. https://escholarship.org/uc/item/55h4h0w7.

Corbane, C., S. K. Lang, S. Pipkins, M. Alleaume, V. E. Deshayes, G. Millán, T. Strasser, J. V. Borre, S. Toon, and M. Förster. 2015. "Remote Sensing for Mapping Natural Habitats and Their Conservation Status–New Opportunities and Challenges." *International Journal of Applied Earth Observation and Geoinformation* 37: 7–16. doi:10.1016/j.jag.2014.11.005.

Cortes, C., and V. Vapnik. 1995. "Support-Vector Networks." *Machine Learning* 20 (3): 273–297. doi:10.1007/BF00994018.

Cutler, D. R., T. C. Edwards, K. H. Beard, A. Cutler, K. T. Hess, J. Gibson, and J. J. Lawler. 2007. "Random Forests for Classification in Ecology." *Ecology* 88 (11): 2783–2792. doi:10.1890/07-0539.1.

Dash, M., and H. Liu. 1997. "Feature Selection for Classification." *Intelligent Data Analysis* 1 (1–4): 131–156. doi:10.1016/S1088-467X(97)00008-5.

Dengler, J., M. Janišová, P. Török, and C. Wellstein. 2014. "Biodiversity of Palaearctic Grasslands: A Synthesis." *Agriculture, Ecosystems & Environment* 182: 1–14. doi:10.1016/j.agee.2013.12.015.

Diaz-Uriarte, R. 2007. "GeneSrF and VarSelRF: A Web-Based Tool and R Package for Gene Selection and Classification Using Random Forest." *BMC Bioinformatics* 8 (1): 328. doi:10.1186/1471-2105-8-328.

Dierschke, H. 1997. "Molinio-Arrhenatheretea (E1)-Kulturgrasland Und Verwandte Vegetationstypen, Teil 1: Arrhenatheretalia-Wiesen Und Weiden Frischer Standorte." *Synopsis Der Pflanzengesellschaften Deutschlands* 3: 1–74.

Dierschke, H., and G. Briemle. 2002. *Kulturgrasland: Wiesen, Weiden Und Verwandte Staudenfluren; 20 Tabellen.* Ulmer: 1–240.

Fassnacht, F. E., F. Hartig, H. Latifi, C. Berger, J. Hernández, P. Corvalán, and B. Koch. 2014. "Importance of Sample Size, Data Type and Prediction Method for Remote Sensing-Based Estimations of Aboveground Forest Biomass." *Remote Sensing of Environment* 154: 102–114. doi:10.1016/j.rse.2014.07.028.

Feilhauer, H., C. Dahlke, D. Doktor, A. Lausch, S. Schmidtlein, G. Schulz, and S. Stenzel. 2014. "Mapping the Local Variability of Natura 2000 Habitats with Remote Sensing." *Applied Vegetation Science* 17 (4): 765–779. doi:10.1111/avsc.12115.

Foerster, S., K. Kaden, M. Foerster, and S. Itzerott. 2012. "Crop Type Mapping Using Spectral–Temporal Profiles and Phenological Information." *Computers and Electronics in Agriculture* 89: 30–40. doi:10.1016/j.compag.2012.07.015.

Förster, M., A. Frick, C. Schuster, and B. Kleinschmit. 2010. "Object-Based Change Detection Analysis for the Monitoring of Habitats in the Framework of the NATURA 2000 Directive with Multi-Temporal Satellite Data." *The International Archives of the Photogrammetry, Remote Sensing and Spatial Information Sciences* 38: 4.

Ghimire, B., J. Rogan, V. R. Galiano, P. Panday, and N. Neeti. 2012. "An Evaluation of Bagging, Boosting, and Random Forests for Land-Cover Classification in Cape Cod, Massachusetts, USA." *GIScience & Remote Sensing* 49 (5): 623–643. doi:10.2747/1548-1603.49.5.623.

Gislason, P. O., J. A. Benediktsson, and J. R. Sveinsson. 2006. "Random Forests for Land Cover Classification." *Pattern Recognition Letters* 27 (4): 294–300. doi:10.1016/j.patrec.2005.08.011.

GRASS Development Team. 2017. *Geographic Resources Analysis Support System (GRASS GIS) Software, Version 7.2.* Open Source Geospatial Foundation. Accessed 1 December 2017. http://grass.osgeo.org.

Laliberte, A. S., E. L. Fredrickson, and A. Rango. 2007. "Combining Decision Trees with Hierarchical Object-Oriented Image Analysis for Mapping Arid Rangelands." *Photogrammetric Engineering & Remote Sensing* 73 (2): 197–207. doi:10.14358/PERS.73.2.197.

Langley, P. 1994. "Selection of Relevant Features in Machine Learning." In *Proceedings of the AAAI Fall Symposium on Relevance*, 184:245–271.

Ma, L., T. Fu, T. Blaschke, M. Li, D. Tiede, Z. Zhou, X. Ma, and D. Chen. 2017. "Evaluation of Feature Selection Methods for Object-Based Land Cover Mapping of Unmanned Aerial Vehicle Imagery Using Random Forest and Support Vector Machine Classifiers." *ISPRS International Journal of Geo-Information*, 6 (2): 51.

Malley, J. D., J. Kruppa, A. Dasgupta, K. G. Malley, and A. Ziegler. 2012. "Probability Machines: Consistent Probability Estimation Using Nonparametric Learning Machines." *Methods of Information in Medicine* 51 (1): 74. doi:10.3414/ME00-01-0052.

Maxwell, A. E., T. A. Warner, and F. Fang. 2018. "Implementation of Machine-Learning Classification in Remote Sensing: An Applied Review." *International Journal of Remote Sensing* 39 (9): 2784–2817. doi:10.1080/01431161.2018.1433343.

Meißner, M., H. Reinecke, S. Herzog, L. Leinen, and G. Brinkmann. 2012. *Vom Wald Ins Offenland: Der Rothirsch Auf Dem Truppenübungsplatz Grafenwöh Raum-Zeit-Verhalten, Lebensraumnutzung, Management*. Aufl. Ahnatal: Frank Fornacon.

Metz, A., A. Schmitt, T. Esch, P. Reinartz, S. Klonus, and M. Ehlers. 2012. "Synergetic Use of TerraSAR-X and Radarsat-2 Time Series Data for Identification and Characterization of Grassland Types-A Case Study in Southern Bavaria, Germany." In *Geoscience and Remote Sensing Symposium (IGARSS)*, 3560–3563. IEEE International. IEEE.

Millard, K., and M. Richardson. 2015. "On the Importance of Training Data Sample Selection in Random Forest Image Classification: A Case Study in Peatland Ecosystem Mapping." *Remote Sensing* 7 (7): 8489–8515. doi:10.3390/rs70708489.

Mountrakis, G., J. Im, and C. Ogole. 2011. "Support Vector Machines in Remote Sensing: A Review." *ISPRS Journal of Photogrammetry and Remote Sensing* 66 (3): 247–259. doi:10.1016/j.isprsjprs.2010.11.001.

Nagendra, H., R. Lucas, J. P. Honrado, R. H. G. Jongman, C. Tarantino, M. Adamo, and P. Mairota. 2013. "Remote Sensing for Conservation Monitoring: Assessing Protected Areas, Habitat Extent, Habitat Condition, Species Diversity, and Threats." *Ecological Indicators* 33: 45–59. doi:10.1016/j.ecolind.2012.09.014.

Nicodemus, K. K., and J. D. Malley. 2009. "Predictor Correlation Impacts Machine Learning Algorithms: Implications for Genomic Studies." *Bioinformatics* 25 (15): 1884–1890. doi:10.1093/bioinformatics/btp331.

Nicodemus, K. K., J. D. Malley, C. Strobl, and A. Ziegler. 2010. "The Behaviour of Random Forest Permutation-Based Variable Importance Measures under Predictor Correlation." *BMC Bioinformatics* 11 (1): 110. doi:10.1186/1471-2105-11-110.

Nitze, I., B. Barrett, and F. Cawkwell. 2015. "Temporal Optimisation of Image Acquisition for Land Cover Classification with Random Forest and MODIS Time-Series." *International Journal of Applied Earth Observation and Geoinformation* 34: 136–146. doi:10.1016/j.jag.2014.08.001.

Pal, M., and G. M. Foody. 2010. "Feature Selection for Classification of Hyperspectral Data by SVM." *IEEE Transactions on Geoscience and Remote Sensing* 48 (5): 2297–2307. doi:10.1109/TGRS.2009.2039484.

Peeters, A., G. Beaufoy, R. M. Canals, A. D. Vliegher, C. Huyghe, J. Isselstein, J. Jones, W. Kessler, D. Kirilovsky, and A.-V. D. P.-V. Dasselaar. 2014. "Grassland Term Definitions and Classifications Adapted to the Diversity of European Grassland-Based Systems." *Grassland Science in Europe* 19: 743–750.

Planet Labs Inc. 2016. "Satellite Imagery Product Specifications." *Satellite Imagery Product Specifications: Version 6.1*: 1–50.

R Core Team. 2017. *R: A Language and Environment for Statistical Computing*. R Foundation for Statistical Computing. Accessed 1 December 2017. https://www.R-project.org/.

Riesch, F., H. G. Stroh, B. Tonn, and J. Isselstein. 2018. "Soil PH and Phosphorus Drive Species Composition and Richness in Semi-Natural Heathlands and Grasslands Unaffected by Twentieth-Century Agricultural Intensification." *Plant Ecology & Diversity* no. just-accepted. doi:10.1080/17550874.2018.1471627.

Robnik-Šikonja, M., P. Savicky, and J. Adeyanju Alao. 2017. *CORElearn: Classification, Regression and Feature Evaluation, R Package Version 1.51.2.* Accessed 1 December 2017. https://CRAN.R-project.org/package=CORElearn.

Rocchini, D., G. M. Foody, H. Nagendra, C. Ricotta, M. Anand, K. S. He, V. Amici, B. Kleinschmit, M. Förster, and S. Schmidtlein. 2013. "Uncertainty in Ecosystem Mapping by Remote Sensing." *Computers & Geosciences* 50: 128–135. doi:10.1016/j.cageo.2012.05.022.

Rodriguez-Galiano, V. F., M. Chica-Olmo, F. Abarca-Hernandez, P. M. Atkinson, and C. Jeganathan. 2012. "Random Forest Classification of Mediterranean Land Cover Using Multi-Seasonal Imagery and Multi-Seasonal Texture." *Remote Sensing of Environment* 121: 93–107. doi:10.1016/j.rse.2011.12.003.

Ruß, G., and A. Brenning. 2010. "Spatial Variable Importance Assessment for Yield Prediction in Precision Agriculture." In *Advances in Intelligent Data Analysis IX*, 184–195. Berlin: Springer. doi:10.1007/978-3-642-13062-5.

Schmidt, J., F. E. Fassnacht, M. Förster, and S. Schmidtlein. 2017. "Synergetic Use of Sentinel-1 and Sentinel-2 for Assessments of Heathland Conservation Status." *Remote Sensing in Ecology and Conservation.* doi:10.1002/rse2.68.

Schmidt, T., C. Schuster, B. Kleinschmit, and M. Förster. 2014. "Evaluating an Intra-Annual Time Series for Grassland Classification—How Many Acquisitions and What Seasonal Origin are Optimal?." *IEEE Journal of Selected Topics in Applied Earth Observations and Remote Sensing* 7 (8): 3428–3439. doi:10.1109/JSTARS.2014.2347203.

Schmidtlein, S., and J. Sassin. 2004. "Mapping of Continuous Floristic Gradients in Grasslands Using Hyperspectral Imagery." *Remote Sensing of Environment* 92 (1): 126–138. doi:10.1016/j.rse.2004.05.004.

Schuster, C., I. Ali, P. Lohmann, A. Frick, M. Förster, and B. Kleinschmit. 2011. "Towards Detecting Swath Events in TerraSAR-X Time Series to Establish NATURA 2000 Grassland Habitat Swath Management as Monitoring Parameter." *Remote Sensing* 3 (7): 1308–1322. doi:10.3390/rs3071308.

Stenzel, S., H. Feilhauer, B. Mack, A. Metz, and S. Schmidtlein. 2014. "Remote Sensing of Scattered Natura 2000 Habitats Using a One-Class Classifier." *International Journal of Applied Earth Observation and Geoinformation* 33: 211–217. doi:10.1016/j.jag.2014.05.012.

Strobl, C., A.-L. Boulesteix, T. Kneib, T. Augustin, and A. Zeileis. 2008. "Conditional Variable Importance for Random Forests." *BMC Bioinformatics* 9 (1): 307. doi:10.1186/1471-2105-9-307.

Strobl, C., A.-L. Boulesteix, A. Zeileis, and T. Hothorn. 2007. "Bias in Random Forest Variable Importance Measures: Illustrations, Sources and a Solution." *BMC Bioinformatics* 8 (1): 25. doi:10.1186/1471-2105-8-25.

Teillet, P. M., B. Guindon, and D. G. Goodenough. 1982. "On the Slope-Aspect Correction of Multispectral Scanner Data." *Canadian Journal of Remote Sensing* 8 (2): 84–106. doi:10.1080/07038992.1982.10855028.

Touw, W. G., J. R. Bayjanov, L. Overmars, L. Backus, J. Boekhorst, M. Wels, and S. A. F. T. Van Hijum. 2012. "Data Mining in the Life Sciences with Random Forest: A Walk in the Park or Lost in the Jungle?." *Briefings in Bioinformatics* 14 (3): 315–326. doi:10.1093/bib/bbs059.

Tyc, G., J. Tulip, D. Schulten, M. Krischke, and M. Oxfort. 2005. "The RapidEye Mission Design." *Acta Astronautica* 56 (1): 213–219. doi:10.1016/j.actaastro.2004.09.029.

Ustuner, M., F. B. Sanli, and S. Abdikan. 2016. "Balanced VS Imbalanced Training Data: Classifying Rapideye Data with Support Vector Machines." In *ISPRS-International Archives of the Photogrammetry, Remote Sensing and Spatial Information Sciences*, 379–384. Prague, Czech Republic. https://doi.org/10.5194/isprs-archives-XLI-B7-379-2016, 2016.

Verikas, A., A. Gelzinis, and M. Bacauskiene. 2011. "Mining Data with Random Forests: A Survey and Results of New Tests." *Pattern Recognition* 44 (2): 330–349. doi:10.1016/j.patcog.2010.08.011.

Vermote, E. F., D. Tanré, J. L. Deuze, M. Herman, and -J.-J. Morcette. 1997. "Second Simulation of the Satellite Signal in the Solar Spectrum, 6S: An Overview." *IEEE Transactions on Geoscience and Remote Sensing* 35 (3): 675–686. doi:10.1109/36.581987.

Wachendorf, M., T. Fricke, and T. Möckel. 2017. *"Remote Sensing as a Tool to Assess Botanical Composition. Structure, Quantity and Quality of Temperate Grasslands."* Grass and Forage Science 73 (1): 1–14. doi: 10.1111/gfs.12312.

Warren, S. D., and R. Büttner. 2008a. "Relationship of Endangered Amphibians to Landscape Disturbance." *Journal of Wildlife Management* 72 (3): 738–744. doi:10.2193/2007-160.

Warren, S. D., and R. Büttner. 2008b. "Active Military Training Areas as Refugia for Disturbance-Dependent Endangered Insects." *Journal of Insect Conservation* 12 (6):671–676. doi:10.1007/s10841-007-9109-2.

Wright, M. N., and A. Ziegler. 2015. "Ranger: A Fast Implementation of Random Forests for High Dimensional Data in C++ and R." *ArXiv Preprint ArXiv:1508.04409* 77 (1). doi:10.18637/jss.v077.i01.

Very high-resolution mapping of emerging biogenic reefs using airborne optical imagery and neural network: the honeycomb worm (*Sabellaria alveolata*) case study

Antoine Collin, Stanislas Dubois, Camille Ramambason and Samuel Etienne

ABSTRACT
Biogenic reefs provide a wide spectrum of ecosystem functions and services, such as biodiversity hotspot, coastal protection, and fishing practices. Honeycomb worm (*Sabellaria alveolata*) reefs, in the Bay of Mont-Saint-Michel (France), constitute the largest intertidal bioconstruction in Europe but undergo anthropogenic pressures (aquaculture-stemmed food/space competition and siltation, fishing-driven trampling). Very high-resolution (VHR) airborne optical data enable cost-efficient biophysical measurements of reef colonies, strongly expected for conservation approaches. A synergy of remotely sensed airborne optical imagery, calibration/validation photoquadrat ground-truth (202/101, respectively), and artificial neural network (ANN) modelling is first used to map *S. alveolata* relative abundance, over the largest bioconstruction in Europe. The best prediction of *S. alveolata* abundance was reached with the infrared–red–green (IRRG) spectral combination and ANN model structured with six neurons ($R^2 = 0.72$, RMSE = 0.08, and $r = 0.85$). The six hyperbolic tangent formulas were applied to the three input spectral bands (IRRG) in order to build six hidden neuronal images, resulting in VHR digital *S. alveolata* abundance model (6547×6566 pixels with 0.5 m pixel size). The innovative model revealed undescribed spatial patterns, namely a reef polarization (perpendicular to the shoreline) of *S. alveolata* abundance: high abundance on forereef and low abundance on backreef.

1. Introduction

Coastal reef builders are able to primarily shape the ecology of local environment through the sediment reworking. By trapping and binding carbonate sands, some cyanobacteria and diatoms produce the stromatolites (Andres and Reid 2006), crustose coralline algae form coatings (Gherardi and Bosence 2001), molluscan vermetidae build bioconstructions (Donnarumma et al. 2017), and cnidarian corals create large barriers (Mumby et al. 2004). Less renown despite their very high productivity, annelids (i.e. worms) can create substantial reefs along tropical and temperate coasts (terebellidae: Degraer et al. 2008; serpulidae: Moore et al. 2009; and sabellariidae: Naylor and Viles 2000).

Honeycomb worm reefs erected by the gregarious tube-building polychaete *Sabellaria alveolata* (Linnaeus, 1767) in the megatidal Bay of Mont-Saint-Michel (BMSM, France) consist of the largest intertidal bioconstruction in Europe (Noernberg et al. 2010; Desroy et al. 2011). Contrary to more common encrusting veneers or hummocks on rocky shores, the Sainte-Anne population in the BMSM develops on soft sediment. It is currently structured as three extensive reef entities within the tidal flats. Such biogenic reefs largely contributes to ecosystem functioning and provide a wide panel of ecosystem services: (1) support, with the significant amount of ecological niches (Dubois, Retière, and Olivier 2002; Jones et al. 2018); (2) regulation, through the sediment stabilization and trapping (Dubois et al. 2005); and (3) culture, by means of recreational shore fisheries (Plicanti et al. 2016). As a biodiversity hotspot and a rare biological and patrimonial heritage, BMSM worm reefs benefit from the European Habitats Directive (Council Directive 92/43/EEC) focusing on the protection and 'Conservation of Natural Habitats' ('biogenic reefs of open seas and tidal areas,' habitat type 1170). However, Sainte-Anne reefs are threatened by local anthropogenic activities, such as seaward intensive Pacific oyster (*Magallana gigas*) and mussel (*Mytilus edulis*) aquaculture, which increases organic and mineral (silt) seston concentration (Dubois, Barillé, and Cognie 2009), interspecific competition for food and space by oyster and mussel that colonize reef surface (Dubois et al. 2006), and shell fishing on reefs, that cause fragmentation by trampling and destructive fishing techniques (Plicanti et al. 2016). Pressure synergy has led to a strong reduction in reef health state between 1970 and 2007, as revealed by diachronic estimates of a spatial 'Reef Health Status Index' (Desroy et al. 2011).

The previous index is the combination of a set of biological (e.g. epibiont covers) and physical (e.g. fragmentation) features of the reef, aimed at quantifying the health (Desroy et al. 2011). A consortium of European coastal scientists, devoted to honeycomb worm reef conservation, indicates that this index is 'a complex time-consuming assessment of the condition of reefs only, that is not widely applicable' (website: honeycombworms.org). Moreover, they advocate that 'the usefulness of ... a generic health index for *S. alveolata* reefs ... should not involve laboratory experimentation, complex measurements or time-consuming processing.' Remote-sensing techniques hold great promises to address this issue given their non-intrusiveness, ease to use, and cost-effectiveness per surface unit (large extent at high resolution). Spaceborne multispectral and hyperspectral, as well as airborne light detection and ranging (lidar) and unmanned vehicle (UAV) imageries were successfully utilized for mapping coral reefs (Collin, Hench, and Planes 2012; Kutser, Miller, and Jupp 2006; Collin et al. 2018a; and Casella et al. 2017) but are still lacking for other reefs, such as annelids, whereas worm reefs were remotely sensed by side-scan sound detection and ranging (Moore et al. 2009; Degraer et al. 2008; Raineault, Trembanis, and Miller 2012; Pearce et al. 2014) and aerial visible (red–green–blue, RGB) photograph interpretation (Brown and Miller 2011; Godet et al. 2011), only two studies used optical data (including infrared, IR). Satellite multispectral imagery (SPOT-4) enabled *S. alveolata* reefs to be mapped at 20 m pixel size (Marchand and Cazoulat 2003), and airborne combination of RGB photointerpretation and IR lidar elevation data at 2 m pixel size were used to delineate *S. alveolata* reefs' extent and estimate their volume (Noernberg et al. 2010). However, the application of both studies into a generic health index is challenging insofar as their spatial resolution is too coarse to account for ecophysiology and reef-building activity. More recent high to very high-resolution (VHR) spaceborne optical data are not yet available over emerged reefs (lying between 2 and 4 m elevation above the national tidal datum epoch), since they are immersed most of the time. The *Ortholittorale* V2 product, collected by the French Ministry for Ecology, Sustainable Development and Energy at low tide, remains, to date, the only VHR optical imagery (0.5 m pixel size) available over the honeycomb worm reefs.

Here we created a method for mapping emerging biogenic reefs at VHR using airborne optical image and selected field data, by focusing on *S. alveolata* relative abundance (Saa). The passive, optical imagery (ranging from IR to blue wavebands at 0.5 m pixel size), acquired from a small aircraft, constitutes the remotely sensed predictors, and an array of RGB photoquadrats (0.5×0.5 m^2) is processed to retrieve the Saa relative abundance, as the ground-truth response. Following comparisons of model performance, the artificial neural network (ANN) is implemented to provide a non-linear regression between both data sets. Our study takes place over Sainte-Anne reefs (Figures 1(a)–(c)), in the heart of BMSM, provided with a maximum tidal range of 14 m. Despite the interest on a single parameter, the Saa open tubes (Figure 1(d)) is deemed as a good proxy for the reef state, given the threats related to silt sedimentation, oyster and mussel colonization, as well as man-made physical degradation. This novel approach has a great potential to contribute to the mapping of worldwide emerging biogenic reefs, aiming at some health indices. Two methodological issues are raised: what are the best spectral predictors? What is the optimized model complexity, featured by the number of neurons? Once the Saa accurately mapped, we examine the spatial patterns of this reef state proxy. Findings are then discussed from the perspective of stakeholders tasked with management of the conservation of intertidal biogenic species adversely affected by anthropogenic activities.

2. Materials and methods

2.1. *Study site*

Sainte-Anne reefs, composed of three adjacent reefs, are situated on the central part (48° 38′ 50″ N, 1° 40′ W) of the megatidal (14 m tidal range) BMSM. Lying between 2 and 4 m elevation (Noernberg et al. 2010) over the French hydrographic zero (i.e. national tidal datum epoch), the Sainte-Anne reefs are spreading over 2.23 km^2 with an estimated volume of 96 301 m^3 (Noernberg et al. 2010). They face massive mussel farming, structured by rows of wooden piles, lying from 0 to 2 m elevation. As the largest intertidal bioconstruction in Europe, the Sainte-Anne reef dynamics can occur in three main morphological shapes (Dubois, Retière, and Olivier 2002): isolated hummocks (ball-shaped structures), coalescent hummocks forming mounds, then platforms. These three stages are modulated by transitional and degraded intermediate stages. Each stage is associated with various sessile species assemblages (*M. gigas*, *M. edulis*, *Crepidula fornicata*, green, brown, and red macroalgae) and specific demographic patterns of *S. alveolata*. Sediment grain-size is essentially composed of gravel, sand, and silt classes.

2.2. *Ground-truth* S. alveolata *abundance (Saa) response*

Fieldwork was carried out on 26 June 2017 using two quadrats (0.5×0.5 m^2, Figure 2(a)), framing RGB photographs, collected with two cameras (Olympus Stylus TG). A series of

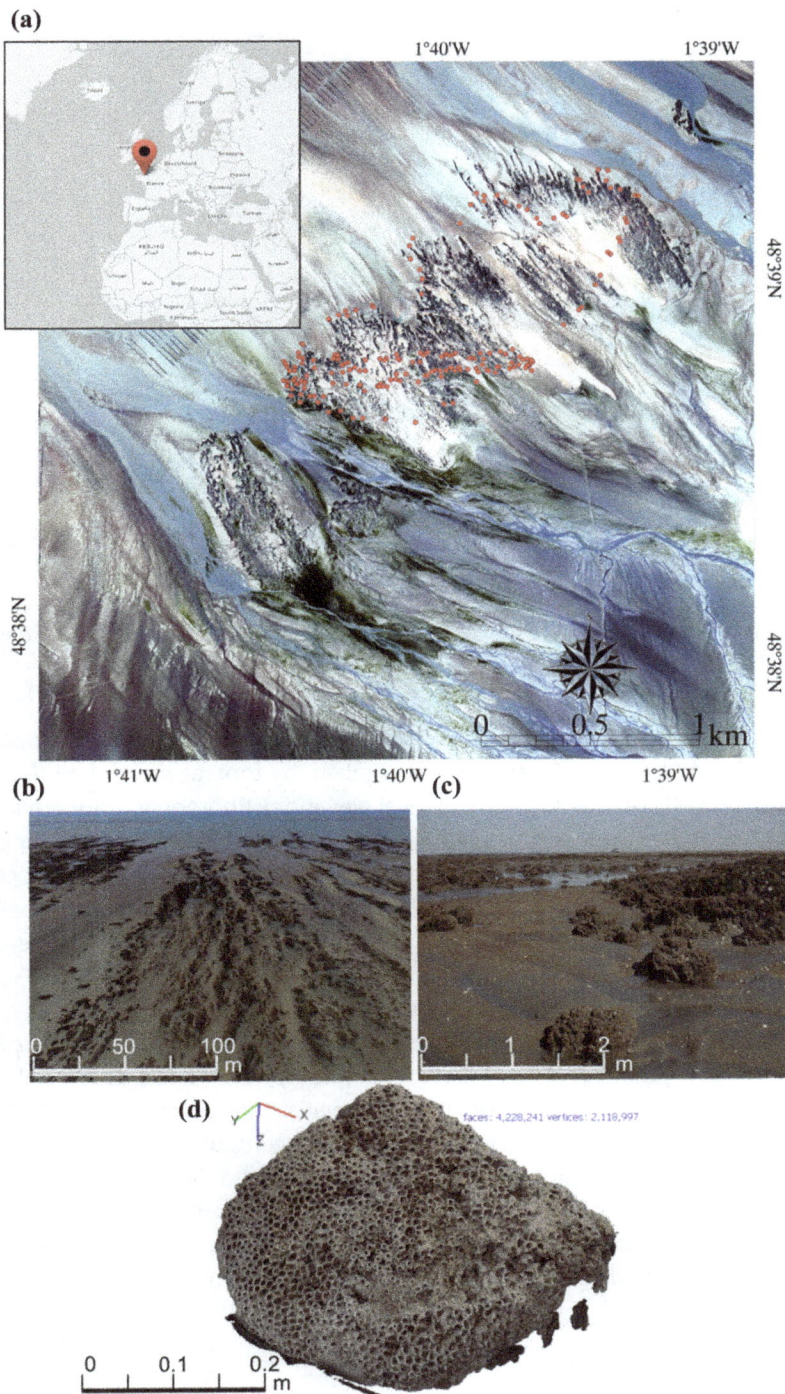

Figure 1. (a) Natural-coloured (red–blue–green) airborne imagery (6547 × 6566 pixels with 0.5 m pixel size) collected on 10 September 2014, over the location of Sainte-Anne three honeycomb worm reefs (*Sabellaria alveolata*), within Bay of Mont-Saint-Michel (Brittany-Normandy, France). Red spots represent photoquadrat locations. (b) Natural-coloured airborne UAV oblique imagery over a portion of the Sainte-Anne reefs. (c) Natural-coloured handborne imagery inside the Sainte-Anne reefs. (d) 3D-model of a honeycomb worm hummock colony draped with natural-coloured imagery.

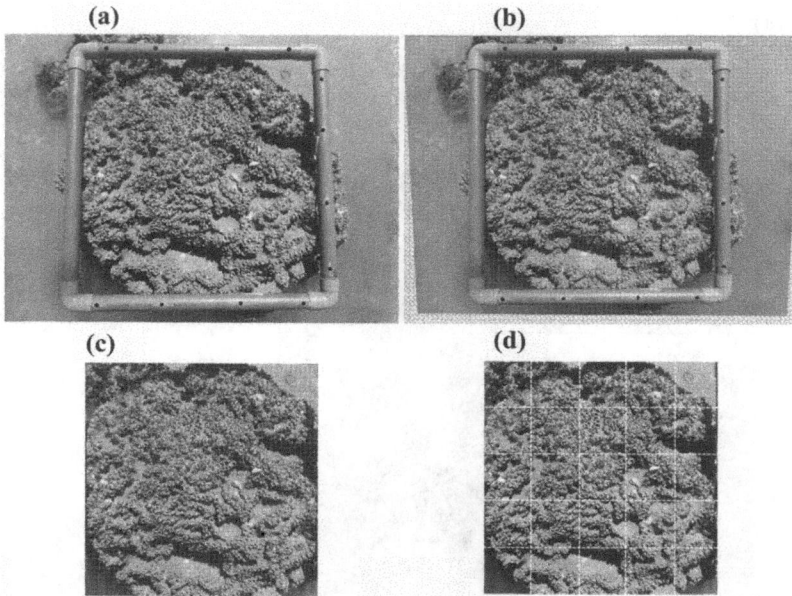

Figure 2. Standardization procedure applied to the (a) original photoquadrat, (b) to correct for the distortion, (c) to crop at the frame scale (0.5 × 0.5 m^2), and (d) to apply a 5 × 5 grid.

303 photoquadrats, geolocated in the WGS84 datum with Global Navigation Satellite System (GNSS) devices (Garmin eTrex®), were taken, by foot, at spring low tide between UTC 13:00 and 15:00 (14:51 – 1.3 m water level elevation). Photoquadrats were sampled to encompass the greatest reef health variability revealed by the most recent mapping work (Rollet et al. 2015). Each photoquadrat was standardized by the following procedure: (1) correction for the geometry acquisition through a distortion method carried out with Photoshop® (Figure 2(b)), (2) cropping the image within the frame (Figure 2(c)), and superimposition of a 5 × 5 grid to analyse independently the 25 image cells (Figure 2(d)).

The standardized photoquadrat enabled the relative abundance (relative area covered by various classes in the 0.25 m^2 plot) of two polychaetes (*S. alveolata* and *Lanice conchilega*), three bivalves (*M. gigas, M. edulis* and *C. fornicata*), fleshy macroalgae, dead bivalves, gravel, sand, slit, and water classes to be quantified. The spatially dominant class in each of the 25 cells 'wins' the cell, and then the relative abundance was computed as the sum of the 25 cells divided by 25. We only exploited the relative abundance of *S. alveolata* (Saa), by means of open tubes' recognition. For the sake of visual interpretation, eight relative abundances of Saa were visually represented by photoquadrats along with their ecological assemblage and reef morphology stage, according to Dubois, Retière, and Olivier (2002) (Table 1).

2.3. *Optical imagery predictors*

The airborne optical survey was conducted on 10 September 2014 (UTC 14:00; 0.36 m water level elevation) using two full frame charge coupled device multispectral cameras: one (UltraCam-Xp, 33 mm focal length) acquiring red, green, and blue wavebands (RGB, Figure 3(a)), and the other one (UltraCam-XpWA, 23 mm focal length), collecting IR, red,

Table 1. Ecological description of the georeferenced photoquadrats (N = 303, 0.5 × 0.5 m^2) from which the abundance of *Sabellaria alveolata* open tubes were retrieved, as a proxy for the honeycomb worm reef state.

Photoquadrat-based class	Ecological assemblage	*Sabellaria alveolata* relative abundance	Worm reef morphology stage (Dubois, Retière, and Olivier 2002)	Colour ramp
	Sand/silt with dead bivalves (shells)	0.0	No *S. alveolata* presence	
	M. gigas/M. edulis/C. fornicata/ fleshy macroalgae/sand/silt/*S. alveolata*	0.1	Degraded isolated *S. alveolata* hummock	
	S. alveolata/sand/silt/*M. gigas/M. edulis/C. fornicata*/fleshy macroalgae	0.2	Isolated *S. alveolata* hummock	
	S. alveolata/sand/silt/shells	0.3	Isolated *S. alveolata* hummock	
	S. alveolata/sand/silt/shells	0.4	Isolated *S. alveolata* hummock	
	S. alveolata/silt/shells	0.5	Coalescent *S. alveolata* hummock	
	S. alveolata/silt/shells	0.6	*S. alveolata* mound	
	S. alveolata	0.7	*S. alveolata* platform	

and green wavebands (infrared–red–green [IRRG], Figure 3(b)). Spectral responses of the four optical wavebands are summarized in Table 2. Analogue image data are recorded at 12 bits, converted to digital numbers at 14 bits, stored without compression at 16 bits, and finally delivered at 8 bits (United States Geological Survey 2010). The freely available *Ortholittorale* V2 product (see hyperlink in 'Acknowledgements' section) has covered all French coastlines, in 2014, with a rigorous acquisition protocol but does not provide spectral specificities of both sensors. The six wavebands, captured with 8 bit radiometric resolution, were orthorectified at

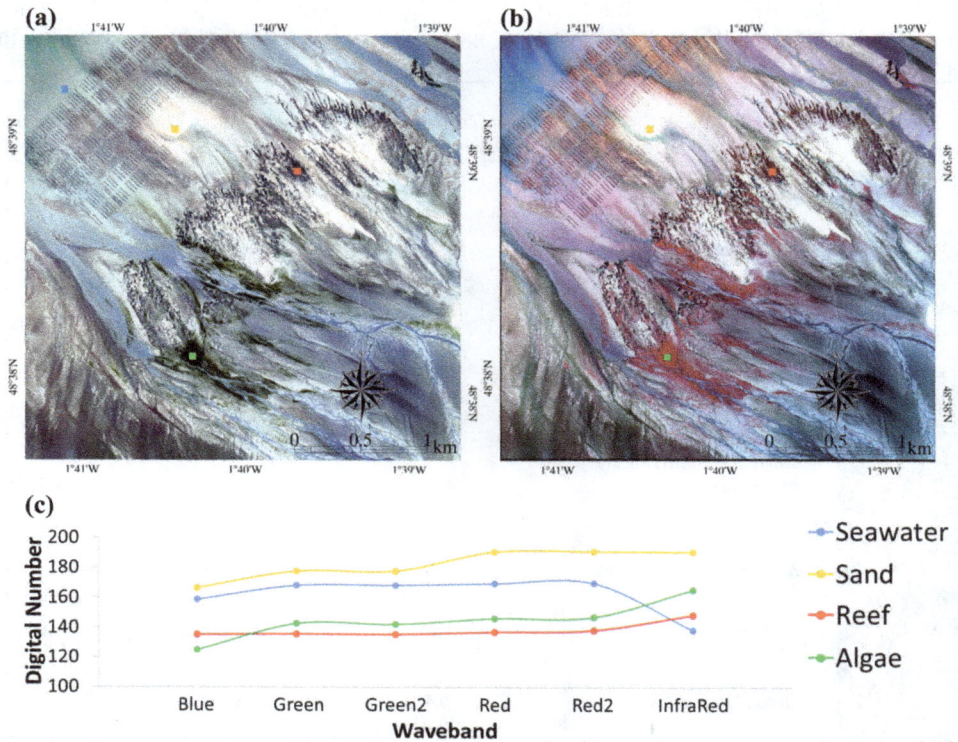

Figure 3. (a) Natural-coloured (red–blue–green) and (b) infrared-coloured (infrared–red–green) air-borne imageries (6547 × 6566 pixels with 0.5 m pixel size) collected on 10 September 2014, over the location of Sainte-Anne reefs. (c) Four rectangles were selected by visual inspection for determining spectral signatures based on *Ortholittorale* V2 of seawater (blue), sand (yellow), reef (red), and algae (green).

Table 2. Spectral sensitivity (in nm) of the airborne optical cameras (UltraCam-Xp and UltraCam-XpWA provided with focal lengths of 33 and 23 mm, respectively).

Blue	Green	Red	Infrared
410–540	480–630	580–700	690–1000

0.5 m spatial resolution in the RGF93 datum (GRS80 spheroid) projected with Lambert93 (conformal conic), the referential French system. Spectral wavebands were highlighted using four spectral signatures associated with four primary features (water, sand, reef, and algae), easily discriminated through image-based inspection (Figure 3(c)).

Insofar as the objective of this study is to target submeter Saa, a thorough registration of coarsely geolocated photoquadrats onto spectral layers was carried out. First, the geographic coordinates of ground-truth were converted into the RGF93 datum, then projected in Lambert93. Second, the converted geolocations were refined by adding the horizontal offset derived from the GNSS measurements and imagery geolocations of eight isolated hummocks, clearly distinctive over imagery. Third, the submeter registration was achieved by translating, where necessary, the refined geolocations onto the correct features using an ultra-high-resolution UAV-stemmed imagery (0.08×0.08 m^2, Collin et al. 2018b).

2.4. *Artificial neural network modelling*

Preliminary comparisons of three main regression learners (ordinary least squares, generalized linear model, and ANN) were carried out, resulting in the ANN selection (Table 3), corroborating another comparison study (Collin, Etienne, and Feunteun 2017). The ANN was selected to develop a robust model to link the discrete Saa response with continuous multispectral predictors.

Based on non-linear modelling, h, the ANN minimizes least squares using a fully connected single-layer perceptron feedforward workflow to predict the Saa response, $h(X)$, from an activated (hyperbolic tangent function, k) sum of the i (ranging from 1 to 7) appropriately weighted, w_i, neurons, n_i, resulting themselves from an appropriate weighting of the multispectral predictors, X (Heermann and Khazenie 1992):

$$h(X) = k\left(\sum_i w_i n_i(X)\right) \qquad (1)$$

Constrained by a single hidden layer, ANN models were developed to test how relevant are the number of neurons, jointly with the implemented spectral combination.

2.5. *Accuracy assessment*

Ground-truth data set was first sorted according to the Saa values and second stratified into 202 calibration and 101 validation subsets (Holdout method) in order to test the acceptability of the modelling. The calibration data set was subject to 1000 computation runs to reach convergent results and therefore avoid stochastic influences, such as the weight initialization. The matching between observed and predicted Saa, based on the validation data set, was quantified using the coefficient of determination (R^2) and the root mean square error (RMSE) of the corresponding linear regression, as well as the corresponding Pearson product-moment correlation coefficient (r).

3. Results

3.1. *Spectral combination and model complexity*

The 101 validation values of Saa were negatively correlated with all spectral wavebands ($r_{Red} = -0.70$, $r_{Green} = -0.76$, $r_{Blue} = -0.74$, $r_{InfraRed} = -0.58$, $r_{Red} = -0.68$, $r_{Green} = -0.74$).

Table 3. Preliminary results of the performance (coefficient of determination, R^2) of three regression models predicting the validation data set of *Sabellaria alveolata* abundance ($N = 101$), in respect to the spectral combination inputs.

Spectral data sets	Ordinary least squares	Generalized linear model (Poisson)	Artificial neural network (3 neurons)
RGB	0.66	0.57	0.69
IRRG	0.65	0.36	0.70
RGB + IRRG	0.63	0.37	0.70

RGB: Red–green–blue; IRRG: infrared–red–green.

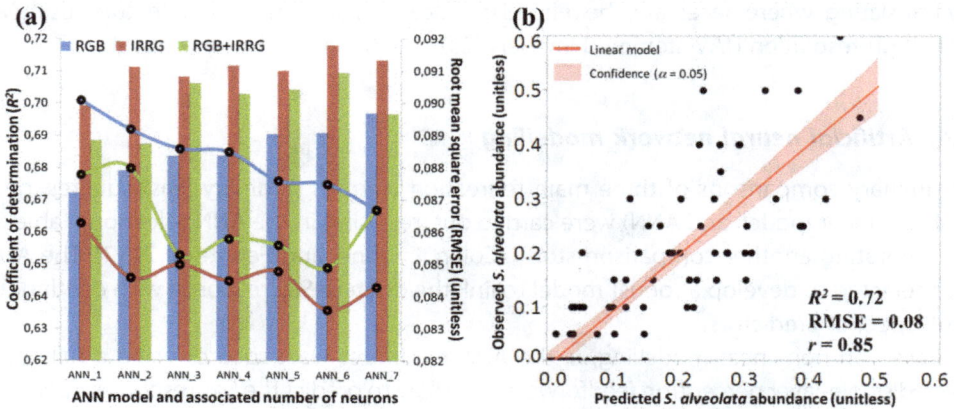

Figure 4. (a) Bar and line plot of the performance (coefficient of determination, R^2, and root mean square error, RMSE, respectively) of the 21 artificial neural network (ANN) models predicting the validation data set of *Sabellaria alveolata* abundance ($N = 101$), as functions of spectral combination inputs and number of hidden neurons. (b) Scatterplot of the validation *versus* predicted *S. alveolata* abundance based on the best ANN model (IRRG as input layers and six neurons within hidden layer).

The influence of the spectral combinations along with the ANN complexity, by interest in the number of neurons in the hidden layer, was tested using the performance metrics of both high R^2 and low RMSE (Figure 4(a)). Overall the agreement between observed and predicted Saa was satisfactory, ranging from a R^2 of 0.67 to 0.72 (Figure 4(a)). The best spectral combination was averagely the IRRG ($R^2 = 0.71$), followed by RGB + IRRG ($R^2 = 0.70$), and finally RGB ($R^2 = 0.68$) (Figure 4(a)). The ANN complexity was optimized with six neurons in the intermediate hidden layer ($R^2 = 0.71$) (Figure 4(a)). The best ANN model was built using the IRRG spectral combination with the six neurons ($R^2 = 0.72$, RMSE = 0.08, and $r = 0.85$, Figure 4(b)). The architecture of the selected ANN model was represented in order to make explicit the doubling number of neurons, compared to the number of input layers (Figure 5).

3.2. *Spatially explicit modelling of* S. alveolata *abundance (Saa)*

The six hyperbolic tangent formulas were applied to the three input spectral bands (IRRG) in order to build six hidden neuronal bands, in turn, implemented into the output linear formula, leading to VHR digital Saa model (Figure 6, 6547 × 6566 pixels with 0.5 m pixel size).

4. Discussion

4.1. *Spectral detection of* S. alveolata *abundance (Saa) and socioecology*

The use of the airborne optical imagery (available over all French metropolitan and most overseas coastal fringes, ≈18,000 km, at low spring tide) has enabled the accurate mapping of the most extended biogenic reef in Europe (i.e. the Sainte-Anne reefs). Enriching the RGB first product of *Ortholittorale* (2000) by IRRG, the second version

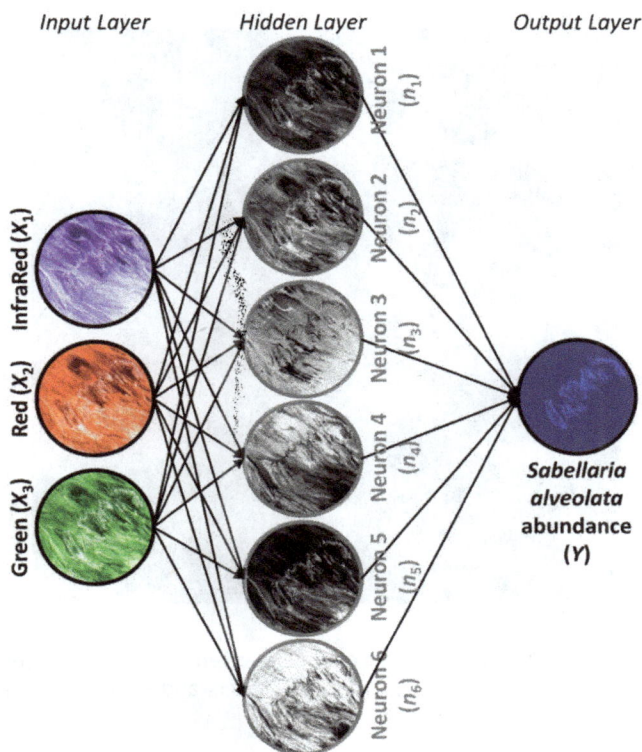

Figure 5. Conceptual flow chart of the artificial neural network modelling based on the infrared, red, and green input layers, the hidden layer provided with six neurons to be able to predict the *Sabellaria alveolata* abundance.

(2014) has leveraged the informational IR waveband. In a broadened context of coastal biogenic reef mapping, some spectral and spatial lessons can be drawn. The interpretation of the visible aerial photography, likely to be constrained by the analyst experience (Brown and Miller 2011; Godet et al. 2011), can be augmented by machine learning procedures (our ANN and cellular automata modelling from Marchand and Cazoulat 2003). Contrary to submerged coral reefs, the emerging biogenic reefs can be better mapped using the IR, strongly absorbed by water and reflected by plant pigment. Enriching the passive RGB data set, IR-derived lidar topography can measure the volume of emerging reefs (Noernberg et al. 2010), even if the lidar IR intensity has not been used yet, contrary to salt marshes (Collin, Long, and Archambault 2010). The integration of the active lidar IR and G backscatters with the passive RGB imagery is strongly advocated to refine the emerging reefs and will be soon possible given the current airborne topo-bathymetric lidar mapping of French coastal fringe (Litto3D® website: diffusion.shom.fr/pro/risques/altimetrie-littorale.html). The 0.5 m resolution encountered in our study outperformed terebellidae and sabellariidae works (100 m in Godet et al. 2011; 75 m in Rollet et al. 2015; Desroy et al. 2011; 20 m in Marchand and Cazoulat 2003; and 2 m in Noernberg et al. 2010).

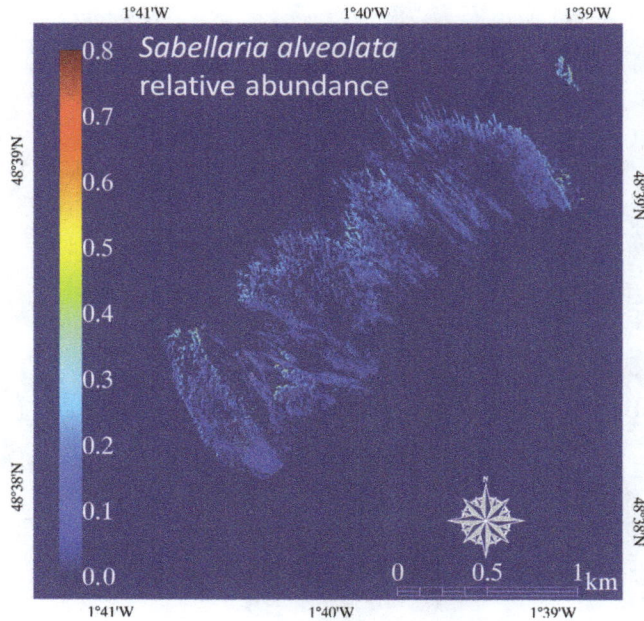

Figure 6. Digital *Sabellaria alveolata* abundance model derived from artificial neural network model with airborne infrared, red, and green bands as input layers and six neurons within hidden layer (6547 × 6566 pixels at 0.5 m pixel size).

Despite the coarse spectral bandwidths, the signature of the reef indicates a low reflection in the visible spectrum, with a slight increase from green (G) to IR (Figure 3(c)). Increasingly negative correlations between Saa and IR, red (R), blue (B), and G show that the reef health proxy might be described by a differential variation occurring between IR and G. Likewise, the normalized difference water index ratioed the G and IR Landsat Thematic Mapper (TM) wavebands (McFeeters 1996). Further spectral investigations, using a portable hyperspectral sensor, are needed to conclude about the key role played by water (moisture) in the reef health mapping. The precise spectral signature of Saa and neighbouring features will also enable a VHR spaceborne proxy to be developed for worm reefs, as successfully done for coral reefs using WorldView-2 imagery (Collin and Planes 2012; Collin, Hench, and Planes 2012). The strong negative correlation between reef health with G might also match the low presence of green macroalgae (e.g. *Ulva* spp.), as highlighted by a long-term survey of the Sainte-Anne reefs health status (Desroy et al. 2011). In this respect, BMSM combines high levels of nutrients, as the junction of landward agricultural runoff and seaward intensive mussel aquaculture, thus favouring the opportunistic seaweed colonization at the expense of *S. alveolata* open tubes. Such a calibrated ANN approach should draw attention to focus on the mapping of green macroalgae. Those fleshy macroalgae have been evidenced not only to affect recruitment patterns (Dubois et al. 2006) but also to potentially contribute to the suspension-feeders' diets, including *S. alveolata* (Lefebvre et al. 2009; Dubois and Colombo 2014). While eutrophication impacts on *S. alveolata* reefs have not been investigated, recent studies emphasized adverse effects on coral reefs (Prouty et al. 2017). In a context of ocean acidification, influence on lower pH on this carbonate-rich reef (made of 60–80% of calcium carbonate grains, Caline et al. 1992) or on biogenic cement polymerization (Fournier, Etienne, and Le Cam 2010) consist of relevant research avenues. Other competitors for space have been targeted, such as farmed *M. gigas* (Desroy et al. 2011), *M. edulis* but also naturally present *Mytilus galloprovincialis* (Jones et al. 2018), whose spatial distribution could importantly explain reef patterns. Based on the occurrence derived from our ground-truth, we greatly recommend taking the invasive gastropod *C. fornicata* mapping into account, due to its trophic competition as a massive population of suspension-feeders.

4.2. *A VHR method to monitor* S. alveolata *abundance (Saa) patterns*

Our spatial modelling has enabled the mapping of Saa at VHR using a reliable method. Fieldworks conducted in April 2015 by Rollet et al. (2015) required 15 persons during 2 days to survey 307 stations using a regular 75 m × 75 m grid mapping (as described in Desroy et al. 2011 for 2001 and 2007 similar survey). Even though such *in situ* studies have led a comprehensive data set (sediment, epifauna, and health status), the spatial scale at stake conspicuously mismatched the fine-scale patterns of *S. alveolata* ecology.

Our outcomes, based on airborne imagery and two persons during 1 day for 303 calibration/validation stations allow reef ecomorphology to be sharply examined, gaining insights into reef responses to exogeneous drivers (Figure 7). The digital Saa model distinctly elucidates a strong polarization of Saa values: highest Saa at the front of reefs, first exposed to sea hydrodynamics and potentially higher coarse sediments and bioclast resuspension (hence increased tube-building activity), contrary

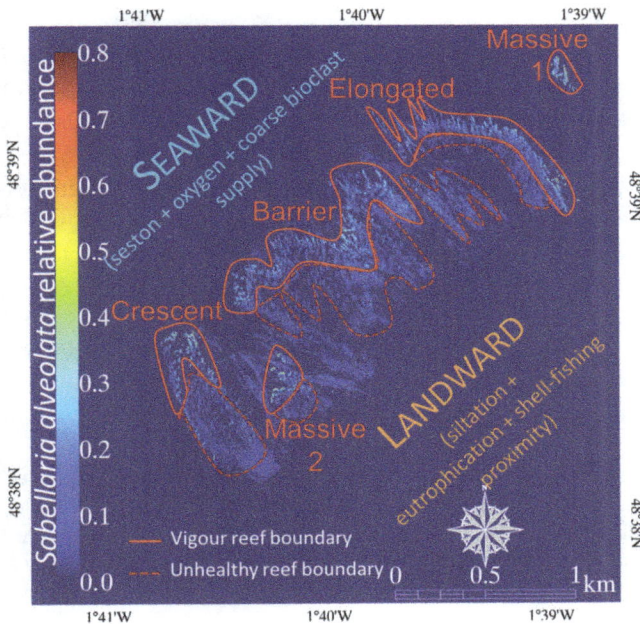

Figure 7. Synthetic conceptual diagram proposing explanation factors of polarized *Sabellaria alveolata* abundance in Sainte-Anne reefs, based on the model derived from artificial neural network model with airborne infrared, red, and green bands as input layers and six neurons within hidden layer (6547 × 6566 pixels at 0.5 m pixel size).

to the back of reefs with lowest Saa, lying on more sheltered and muddier environments (Bonnot-Courtois et al. 2008). The back-reef is subject to finer grained resources, which hamper filtration activity as tentacular filaments of *S. alveolata* are clogged and gut contents are more rapidly filled by poor-quality suspended food sources due to an increase in inorganic content (Dubois et al. 2005). Moreover, apparently unhealthy state of the back-reef might be due to the higher occurrences of oysters, hence leading to higher trampling and destructive shell fishing techniques (Plicanti et al. 2016). In addition to the characterization of the reef polarization perpendicular to the shoreline, the VHR Saa mapping feature five separated reefs composing Sainte-Anne (unlike the three main parts identified in Desroy et al. (2011): a central front barrier reef with a vigorous front core and northern small seaward elongations, a northern crescent reef with developed seaward elongations, a southern crescent reef, two massive intermediate (between sea and land) coalescent reefs. Further examinations of landscape connectivity (using dedicated software, such as Graphab) may facilitate modelling of ecological networks and ultimately help stakeholders to include biodiversity conservation into coastal spatial planning.

5. Conclusion

This novel research shows that airborne optical imagery, ranging from green to IR, brings enough information to robustly map emerging biogenic reefs at VHR. The original findings derived from the largest bioconstruction in Europe (honeycomb worm reefs) can be summarized as follows.

(1) *S. alveolata* relative abundance (Saa) of emerging reefs can be fully surveyed by airborne RGB and/or IRRG cameras at the colony-scale (0.5 m) during low spring tide.
(2) IRRG are better predictors of Saa than RGB (R^2 = 0.71 and 0.68, respectively).
(3) Adding RGB to IRRG reduce the prediction performance of Saa (R^2 = 0.70).
(4) ANN, as a robust non-linear model, is optimized with a hidden layer provided with six neurons in order to predict Saa (R^2 = 0.71).
(5) The best prediction of Saa was reached with the IRRG spectral combination and ANN model structured with six neurons (R^2 = 0.72, RMSE = 0.08, and r = 0.85).

Acknowledgements

Authors gratefully thank French Minister for Ecology, Sustainable Development and Energy for the airborne multispectral acquisition and dissemination related to *Ortholittorale* V2 product (http://cartelie.application.developpement-durable.gouv.fr/cartelie/voir.do?carte=telecharg_ol_v2_I93&service=CEREMA). Hélène Gloria and Dorothée James are also greatly acknowledged for their fieldwork involvement.

Disclosure statement

No potential conflict of interest was reported by the authors.

References

Andres, M. S., and R. P. Reid. 2006. "Growth Morphologies of Modern Marine Stromatolites: A Case Study from Highborne Cay, Bahamas." *Sedimentary Geology* 185 (3–4): 319–328. doi:10.1016/j.sedgeo.2005.12.020.

Bonnot-Courtois, C., P. Bassoullet, B. Tessier, F. Cayocca, P. Le Hir, and A. Baltzer. 2008. "Remaniements Sédimentaires Superficiels Sur L'estran Occidental De La Baie Du Mont-Saint-Michel." *European Journal of Environmental and Civil Engineering* 12: 51–65. doi:10.1080/19648189.2008.9692995.

Brown, J. R., and D. C. Miller. 2011. "Persistence and Distribution of Temperate Intertidal Worm Reefs in Delaware Bay: A Comparison of Biological and Physical Factors." *Estuaries and Coasts* 34 (3): 583–596. doi:10.1007/s12237-011-9387-5.

Caline, B., Y. Gruet, C. Legendre, J. Le Rhun, A. L'Homer, R. Mathieu, and R. Zbinden. 1992. *The Sabellariid Reefs in the Bay of Mont Saint-Michel, France. Ecology, Geomorphology, Sedimentology and Geologic Implications*. Edited translated by D. W. Kirtley Stuart, Florida: Florida Oceanographic Society, Contributions to Marine Science 1. 156 p.

Casella, E., A. Collin, D. Harris, S. Ferse, S. Bejarano, V. Parravicini, J. L. Hench, and A. Rovere. 2017. "Mapping Coral Reefs Using Consumer-Grade Drones and Structure from Motion Photogrammetry Techniques." *Coral Reefs (Online)* 36 (1): 269–275. doi:10.1007/s00338-016-1522-0.

Collin, A., S. Dubois, D. James, C. Ramambason, H. Gloria, E. Feunteun, and S. Etienne. 2018b. "Complexité Structurale Des Récifs Biogéniques D'hermelles Par Drone Aérien." Proceedings of the 2nd merIGéo, Aix-en-Provence, 23–26.

Collin, A., S. Etienne, and E. Feunteun. 2017. "VHR Coastal Bathymetry Using WorldView-3: Colour versus Learner." *Remote Sensing Letters* 8 (11): 1072–1081. doi:10.1080/2150704X.2017.1354261.

Collin, A., J. L. Hench, and S. Planes. 2012. "A Novel Spaceborne Proxy for Mapping Coral Cover." In *Proceedings of the 12th International Coral Reef Symposium*, Cairns, 1–5.

Collin, A., B. Long, and P. Archambault. 2010. "Salt-Marsh Characterization, Zonation Assessment and Mapping through a Dual-Wavelength LiDAR." *Remote Sensing of Environment* 114 (3): 520–530. doi:10.1016/j.rse.2009.10.011.

Collin, A., and S. Planes. 2012. "Enhancing Coral Health Detection Using Spectral Diversity Indices from Worldview-2 Imagery and Machine Learners." *Remote Sensing* 4 (10): 3244–3264. doi:10.3390/rs4103244.

Collin, A., C. Ramambason, Y. Pastol, E. Casella, A. Rovere, L. Thiault, B. Espiau, et al. 2018a. "Very High Resolution Mapping of Coral Reef State Using Airborne Bathymetric LiDAR Surface-Intensity Predictors. visible drone response, and neural network". International Journal Remote Sensing. Re-submitted.

Degraer, S., G. Moerkerke, M. Rabaut, G. Van Hoey, I. Du Four, M. Vincxand, and J.-P. Van Lancker. 2008. "Very-High Resolution Side-Scan Sonar Mapping of Biogenic Reefs of the Tube-Worm Lanice Conchilega." *Remote Sensing of Environment* 112 (8): 3323–3328. doi:10.1016/j.rse.2007.12.012.

Desroy, N., S. Dubois, J. Fournier, L. Ricquiers, P. Le Mao, L. Guérin, and A. Legendre. 2011. "The Conservation Status of Sabellaria Alveolata (L.)(Polychaeta: Sabellariidae) Reefs in the Bay of Mont-Saint-Michel." *Aquatic Conservation: Marine and Freshwater Ecosystems* 21 (5): 462–471. doi:10.1002/aqc.1206.

Donnarumma, L., R. Sandulli, L. Appolloni, F. Di Stefano, and G. F. Russo. 2017. "Morpho-Structural and Ecological Features of a Shallow Vermetid Bioconstruction in the Tyrrhenian Sea (Mediterranean Sea, Italy)." *Journal of Sea Research* 131: 61–68.

Dubois, S., L. Barillé, and B. Cognie. 2009. "Feeding Response of the Polychaete Sabellaria Alveolata (Sabellariidae) to Changes in Seston Concentration." *Journal of Experimental Marine Biology and Ecology* 376 (2): 94–101. doi:10.1016/j.jembe.2009.06.017.

Dubois, S., L. Barillé, B. Cognie, and P. G. Beninger. 2005. "Particle Capture and Processing Mechanisms in Sabellaria Alveolata (Polychaeta: Sabellariidae)." *Marine Ecology Progress Series* 301: 159–171. doi:10.3354/meps301159.

Dubois, S., J. A. Commito, F. Olivier, and C. Retière. 2006. "Effects of Epibionts on Sabellaria Alveolata (L.) Biogenic Reefs and Their Associated Fauna in the Bay of Mont Saint-Michel." *Estuarine, Coastal and Shelf Science* 68 (3–4): 635–646. doi:10.1016/j.ecss.2006.03.010.

Dubois, S., C. Retière, and F. Olivier. 2002. "Biodiversity Associated with Sabellaria Alveolata (Polychaeta: Sabellariidae) Reefs: Effects of Human Disturbances." *Journal of the Marine Biological Association of the United Kingdom* 82 (5): 817–826. doi:10.1017/S0025315402006185.

Dubois, S. F., and F. Colombo. 2014. "How Picky Can You Be? Temporal Variations in Trophic Niches of Co-Occurring Suspension-Feeding Species." *Food Webs* 1: 1–9. doi:10.1016/j.fooweb.2014.07.001.

Fournier, J., S. Etienne, and J. B. Le Cam. 2010. "Inter- and Intraspecific Variability in the Chemical Composition of the Mineral Phase of Cements from Several Tube-Building Polychaetes." *Geobios* 43: 191–200. doi:10.1016/j.geobios.2009.10.004.

Gherardi, D. F. M., and D. W. J. Bosence. 2001. "Composition and Community Structure of the Coralline Algal Reefs from Atol Das Rocas, South Atlantic, Brazil." *Coral Reefs (Online)* 19 (3): 205–219. doi:10.1007/s003380000100.

Godet, L., J. Fournier, M. Jaffré, and N. Desroy. 2011. "Influence of Stability and Fragmentation of a Worm-Reef on Benthic Macrofauna." *Estuarine, Coastal and Shelf Science* 92 (3): 472–479. doi:10.1016/j.ecss.2011.02.003.

Heermann, P. D., and N. Khazenie. 1992. "Classification of Multispectral Remote Sensing Data Using a Back-Propagation Neural Network." *IEEE Transactions on Geoscience and Remote Sensing* 30 (1): 81–88. doi:10.1109/36.124218.

Jones, A. G., S. F. Dubois, N. Desroy, and J. Fournier. 2018. "Interplay between Abiotic Factors and Species Assemblages Mediated by the Ecosystem Engineer Sabellaria Alveolata (Annelida: Polychaeta)." *Estuarine Coastal and Shelf Science* 200: 1–18. doi:10.1016/j.ecss.2017.10.001.

Kutser, T., I. Miller, and D. L. Jupp. 2006. "Mapping Coral Reef Benthic Substrates Using Hyperspectral Space-Borne Images and Spectral Libraries." *Estuarine, Coastal and Shelf Science* 70 (3): 449–460. doi:10.1016/j.ecss.2006.06.026.

Lefebvre, S., J. C. Marin-Leal, S. Dubois, F. Orvain, J. L. Blin, M. P. Bataille, A. Ourry, and R. Galois. 2009. "Seasonal Dynamics of Trophic Relationships among Co-Occurring Suspension-Feeders in Two Shellfish-Culture Dominated Ecosystems." *Estuarine Coastal and Shelf Science* 82: 415–425. doi:10.1016/j.ecss.2009.02.002.

Marchand, Y., and R. Cazoulat. 2003. "Biological Reef Survey Using Spot Satellite Data Classification by Cellular Automata Method - Bay of Mont Saint-Michel (France)." *Computers & Geosciences* 29: 413–421. doi:10.1016/S0098-3004(02)00116-4.

McFeeters, S. K. 1996. "The Use of the Normalized Difference Water Index (NDWI) in the Delineation of Open Water Features." *International Journal of Remote Sensing* 17: 1425–1432. doi:10.1080/01431169608948714.

Moore, C. G., C. Richard Bates, J. M. Mair, G. R. Saunders, D. B. Harries, and A. R. Lyndon. 2009. "Mapping Serpulid Worm Reefs (Polychaeta: Serpulidae) for Conservation Management." *Aquatic Conservation: Marine and Freshwater Ecosystems* 19 (2): 226–236. doi:10.1002/aqc.v19:2.

Mumby, P. J., W. Skirving, A. E. Strong, J. T. Hardy, E. F. LeDrew, E. J. Hochberg, R. P. Stumpf, and L. T. David. 2004. "Remote Sensing of Coral Reefs and Their Physical Environment." *Marine Pollution Bulletin* 48 (3–4): 219–228. doi:10.1016/j.marpolbul.2003.10.031.

Naylor, L. A., and H. A. Viles. 2000. "A Temperate Reef Builder: An Evaluation of the Growth, Morphology and Composition of Sabellaria Alveolata (L.) Colonies on Carbonate Platforms in South Wales." *Geological Society, London, Special Publications* 178 (1): 9–19. doi:10.1144/GSL.SP.2000.178.01.02.

Noernberg, M. A., J. Fournier, S. Dubois, and J. Populus. 2010. "Using Airborne Laser Altimetry to Estimate Sabellaria Alveolata (Polychaeta: Sabellariidae) Reefs Volume in Tidal Flat Environments." *Estuarine, Coastal and Shelf Science* 90 (2): 93–102. doi:10.1016/j.ecss.2010.07.014.

Pearce, B., J. M. Fariñas-Franco, C. Wilson, J. Pitts, A. deBurgh, and P. J. Somerfield. 2014. "Repeated Mapping of Reefs Constructed by Sabellaria Spinulosa Leuckart 1849 at an Offshore Wind Farm Site." *Continental Shelf Research* 83: 3–13. doi:10.1016/j.csr.2014.02.003.

Plicanti, A., R. Domínguez, S. Dubois, and I. Bertocci. 2016. "Human Impacts on Biogenic Habitats: Effects of Experimental Trampling on Sabellaria Alveolata (Linnaeus, 1767) Reefs." *Journal of Experimental Marine Biology and Ecology* 478: 34–44. doi:10.1016/j.jembe.2016.02.001.

Prouty, N. G., A. Cohen, K. K. Yates, C. D. Storlazzi, P. W. Swarzenski, and D. White. 2017. "Vulnerability of Coral Reefs to Bioerosion from Land-Based Sources of Pollution." *Journal of Geophysical Research: Oceans*122: 9319–9331.doi:10.1002/2017JC013264.

Raineault, N. A., A. C. Trembanis, and D. C. Miller. 2012. "Mapping Benthic Habitats in Delaware Bay and the Coastal Atlantic: Acoustic Techniques Provide Greater Coverage and High Resolution in Complex, Shallow-Water Environments." *Estuaries and Coasts* 35 (2): 682–699. doi:10.1007/s12237-011-9457-8.

Rollet, C., D. Mathérion, N. Desroy, and P. Le Mao. 2015. "Suivi De L'état De Conservation Des Récifs D'hermelles (*Sabellaria Alveolata*)". Rapport final, décembre 2015, Ifremer/ODE/LITTORAL/LER/BN-15-008, Projet Life 12 ENV/FR/316 – Expérimentation pour une gestion durable et concertée de la pêche à pied de loisir – LIFE+ Pêche à pied de loisir, 32 p. + annexes.

United States Geological Survey. 2010. *Digital Aerial Sensor Certification Report for the Microsoft Vexcel UltraCamD, UltraCamX, UltraCamXp, and UltraCamXp WA Models*. Sioux Falls, South Dakota: Department of the Interior US Geological Survey. 19 p.

Very high resolution mapping of coral reef state using airborne bathymetric LiDAR surface-intensity and drone imagery

Antoine Collin, Camille Ramambason, Yves Pastol, Elisa Casella, Alessio Rovere, Lauric Thiault, Benoît Espiau, Gilles Siu, Franck Lerouvreur, Nao Nakamura, James L. Hench, Russell J. Schmitt, Sally J. Holbrook, Matthias Troyer and Neil Davies

ABSTRACT

Very high resolution (VHR) airborne data enable detection and physical measurements of individual coral reef colonies. The bathymetric LiDAR system, as an active remote sensing technique, accurately computes the coral reef ecosystem's surface and reflectance using a single green wavelength at the decimetre scale over 1-to-100 km^2 areas. A passive multispectral camera mounted on an airborne drone can build a blue-green-red (BGR) orthorectified mosaic at the centimetre scale over 0.01-to-0.1 km^2 areas. A combination of these technologies is used for the first time here to map coral reef ecological state at the submeter scale. Airborne drone BGR values (0.03 m pixel size) serve to calibrate airborne bathymetric LiDAR surface and intensity data (0.5 m pixel size). A classification of five ecological states is then mapped through an artificial neural network (ANN). The classification was developed over a small area (0.01 km^2) in the lagoon of Moorea Island (French Polynesia) at VHR (0.5 m pixel size) and then extended to the whole lagoon (46.83 km^2). The ANN was first calibrated with 275 samples to determine the class of coral state through LiDAR-based predictors; then, the classification was validated through 135 samples, reaching a satisfactory performance (overall accuracy = 0.75).

1. Introduction

Coral reefs host 25% of the marine biodiversity but are increasingly subject to global ocean-climate changes and local anthropogenic activities. Fine-scale monitoring of coral reef ecosystems and associated ecosystem services are needed for their management and spatial planning. Coral reef mapping usually relies on remote sensing for cost-effectively identifying their structural complexity, benthic composition, and regime surrogates over large areas (Goodman, Samuel, and Stuart 2013; Hedley et al. 2016). Spaceborne multispectral imagery demonstrates great spatial potential to accurately map coral reef colonies (Collin, Hench, and Planes 2012), habitats (Collin et al. 2016), health (Collin and Planes 2012; Collin, Archambault, and Planes 2014) and resilience (Rowlands et al. 2012; Knudby et al. 2013; Collin, Nadaoka, and Bernardo 2015). Airborne passive hyperspectral imagery, provided with dozens of spectral bands, enables coral reef benthos, substrates, and bathymetry to be significantly improved (Leiper et al. 2014). Airborne (usually on manned aircraft) active light detection and ranging (LiDAR) is now the reference system for measuring bathymetry, outperforming waterborne sound detection and ranging (SoNAR) devices, which are strongly impeded by shallow features, specifically in the coastal realm where coral reefs thrive (Costa, Battista, and Pittman 2009). LiDAR-derived morphometry indices can reveal efficient proxies for ecosystem characteristics, for example, estimates of reef fish assemblages (Wedding et al. 2008). Yet despite the increase in discrimination power showed over benthic habitats bathed with turbid waters, LiDAR indices have not been used to date to exploit the spectral information associated with water-penetrating green LiDAR wavelength for coral reef monitoring (Collin, Archambault, and Long 2008, 2011; Collin, Long, and Archambault 2011).

Unmanned airborne vehicles (UAVs or simply 'drones') are becoming an integral component of the scientific toolbox for coral reef research and management. Equipped with blue-green-red (BGR) spectral cameras, drones are able to measure coral reef bathymetry and derived terrain roughness at very high resolution (VHR) using the photogrammetry approach (Leon et al. 2015; Casella et al. 2017). The 3D point cloud, permitting 2D orthorectified BGR mosaics and 2.5D digital surface models (DSM), results from the multi-angle information of a single scene made possible by the spatially-even acquisition of BGR imagery from a moving airborne drone flying at low altitude (from 30 to 150 m): so-called 'structure-from-motion'. The images and by-products yield spatial resolution at the centimetre scale (i.e. 0.03 m pixel size). Coral reef states can be significantly distinguished using the resulting 0.03 m BGR orthomosaic drone dataset, enabling classification of reef ecological states.

Here we describe a methodology for creating the first coral reef ecological state map at VHR based solely on regional airborne LiDAR 'predictors' trained with local 'response' imagery from the drone. The bathymetric LiDAR Riegl VQ-820-G, mounted on a small plane or helicopter, serves as the remotely sensed 1-to-100 km^2 predictors with four measurements of surface and intensity (green) per m^2. The BGR GoPro, mounted on a consumer-grade airborne drone (DJI phantom 2), is used as the remotely sensed 0.01-to-0.1 km^2 response. Spearheading machine learners in satellite-based coastal prediction (Collin, Etienne, and Feunteun 2017), an artificial neural network (ANN) classifier is developed to provide a robust, yet simple, algorithm linking the two datasets. Our study takes place on one of the best-studied islands in the world (Cressey 2015): Moorea (French Polynesia, Figure 1), a volcanic island with fringing, barrier and outer coral reefs in the central South Pacific Ocean. It contributes to efforts to build a 4D model—an Island Digital Ecosystem Avatar (IDEA)—of Moorea and to simulate of future states of the social-ecological system in support of scenario-based planning (Davies et al. 2016). We follow a drone-based assessment of ecological state (coral reef state classification; Table 1) and combine it with LiDAR-based data to spatially classify the coral reef state at VHR over a small area and then extend this to the whole island. Findings are discussed with a view to how this approach could advance an automated workflow for coral reef mapping.

2. Materials and methods

2.1. *Study site*

The study site is located in the northern lagoon of Moorea Island (17°33′S, 149°50′W) in the Society Archipelago (French Polynesia, Figure 1(a)). Moorea demonstrates a highly resilient coral reefs (Adjeroud et al. 2009), especially its outer slope, which following the extremely low coral cover (2%) due to 2007–2010 outbreak of corallivore crown-of-thorns sea star (*Acanthaster planci*) and 2010 Oli cyclone strike, is recovering to record rates close to 70% (Chancerelle, pers. comm.). Located inside the 46.83 km^2 Moorea lagoon, the study site covers 11 710 m^2 with a maximum depth of 2 m. It is bathed in oligotrophic, thus clear, seawater including various taxa of reef-building corals (*Porites, Acropora, Pocillopora, Montipora*), red calcareous algae (*Lithothamnium*), fleshy algae (red, brown and green) and a diversity of geomorphic features (rubble, sand and pavement).

2.2. *Drone visible response*

A drone-based spectral survey (Figure 1(b)) was carried out on 17 August 2015 using a BGR camera (GoPro Hero 4) mounted on a consumer-grade drone (DJI Phantom 2). Calm sea and low sun elevation angle were optimal conditions for this survey. A series of 360 geolocated BGR photographs, acquired at 30 m altitude at nadir, were mosaicked then processed using the photogrammetry software Agisoft Photoscan (http://www.agisoft. ru). Constrained by nine ground control points and three scale bars, the resulting orthorectified mosaic (WGS 84 datum and UTM 6S projection) has 0.03 m resolution (see Casella et al. 2017 for further details) and was therefore deemed as precise enough to be used as air-truth (Figure 1(c), Collin, Lambert, and Etienne 2018). A total of 410 sampling points over the BGR orthomosaic, corresponding to as many LiDAR soundings, were visually interpreted by an expert and classified into five ecological states (Figure 2 (a) and Table 1), each one composed of 55 training and 27 validation sub-datasets.

2.3. *Lidar surface and intensity predictors*

The airborne LiDAR campaign was conducted from 10 to 26 June 2015 (one month before the drone flight) using a Riegl VQ-820-G hydrographic laser scanner mounted on a small plane. The sensor was operated at 251 kHz, providing a minimum sounding density of four points per m^2 (0.5 m) and vertical accuracy of 0.15 m, computed from

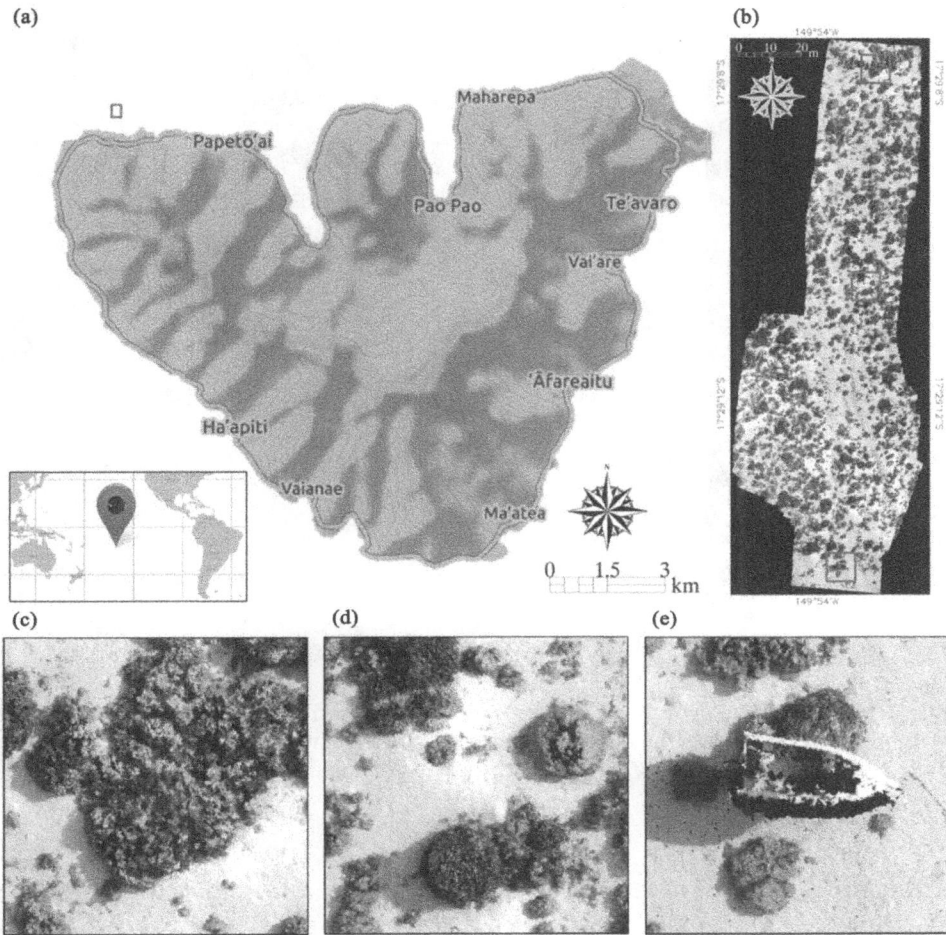

Figure 1. (a) Moorea Island (French Polynesia) was surveyed by bathymetric LiDAR at island scale (10–26 June 2015) and over a small study area by airborne drone (17 August 2015; red rectangle). (b) Natural-coloured (blue-green-red) drone survey provides spectral information at 0.03 m pixel size (2133 × 6095 pixels), enabling resolution of coral reef (c) assembled colonies, (d) single colonies on sand/pavement, or (e) anthropogenic features.

43,798 comparisons (Pastol, Chamberlain, and Sinclair 2016). This bathymetric LiDAR pulses an electromagnetic radiation (532 nm wavelength, namely green) from the aircraft and records its travel time in air and water by means of a waveform (Collin, Archambault, and Long 2008). LiDAR surface and intensity are computed on-the-fly for each sounding by converting the time between sea surface and bottom green echoes into distance (knowing the light speed into water), and by recording the peak of bottom green echo, respectively. Maximum depth ever recorded by bathymetric LiDAR reached 76.1 m in Moorea Island during the studied survey (Pastol, Chamberlain, and Sinclair 2016) given the water clarity due to oligotrophic waters. Since our Moorea study limits to the shallow waters (< 10 m depth), LiDAR intensity has been directly processed with no water correction. As each LiDAR surface and intensity sounding is duly located by the

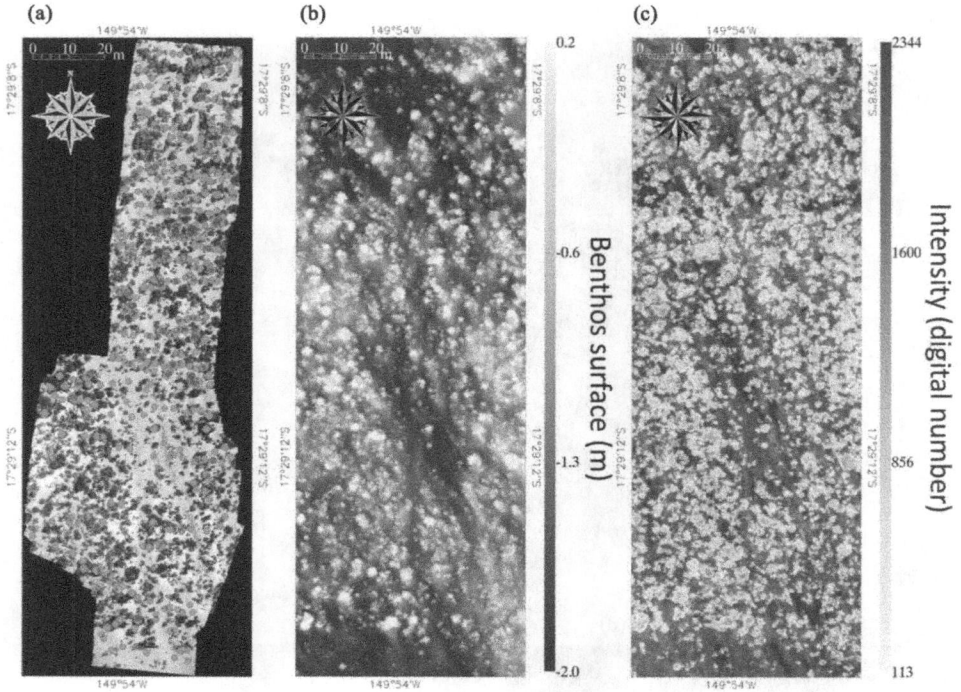

Figure 2. Maps of the (a) natural-coloured imagery with 410 air-truth sampling sites (red transparent disks), (b) bathymetric LiDAR surface soundings, and (c) bathymetric LiDAR intensity (532 nm wavelength) soundings. (a) is at 0.03 m, whereas (b) and (c) are at 0.5 m pixel size.

combination of HR global navigation satellite system and inertial measurement unit, digital surface and intensity models (DSM and DIM, Figure 2(b, c)) can be calculated using ordinary kriging method applied to LiDAR sounding clouds. LiDAR points and rasters were geographically referenced to WGS84 UTM 6S and altimetrically zeroed as the mean sea level (SHOM 2016). Drone-derived imagery was registered with LiDAR data using a 1st degree polynomial function and resampled with cubic convolution.

2.4. Artificial neural network classification

Given their performance in a comparative analysis (Collin, Etienne, and Feunteun 2017), we use an ANN approach as a classification procedure binding the drone-based air-truth and LiDAR-based variables.

The ANN builds non-linear classifications by minimizing least squares using a multi-layer perceptron classifying ecological state response, $h(X)$ (Table 1) with the LiDAR surface and intensity predictors, X, through a constant, k, and intermediate weighted, w_i, functions called neurons, n_i (Heermann and Khazenie 1992):

$$h(X) = k\left(\sum_i w_i n_i(X)\right) \tag{1}$$

Table 1. Ecological description of the five coral reefscape states identified on airborne drone blue-green-red imagery (0.03 m spatial resolution) enabling a coral reef state classification to be created and colour-coded.

Drone-based state					
Ecological composition	*Acropora/ Pocillopora/ Montipora* stony corals	*Acropora/ Pocillopora/ Montipora* stony corals with red calcareous algae	*Porites* stony corals	Microalgae on rubble	Sand on pavement
Structural complexity	Very High roughness	High roughness	Medium roughness	Low roughness	Very low roughness
Coral reef state	1	2	3	4	5
Colour class					

Neurons n_i are hereinafter based on hyperbolic tangents. ANN constrained by a single hidden layer provided with two neurons so the number of neurons to be in synergy with the number of inputs (predictors, Figure 3). Trained by the 275 calibration samples, the ANN will be validated by the remaining 135 validation samples.

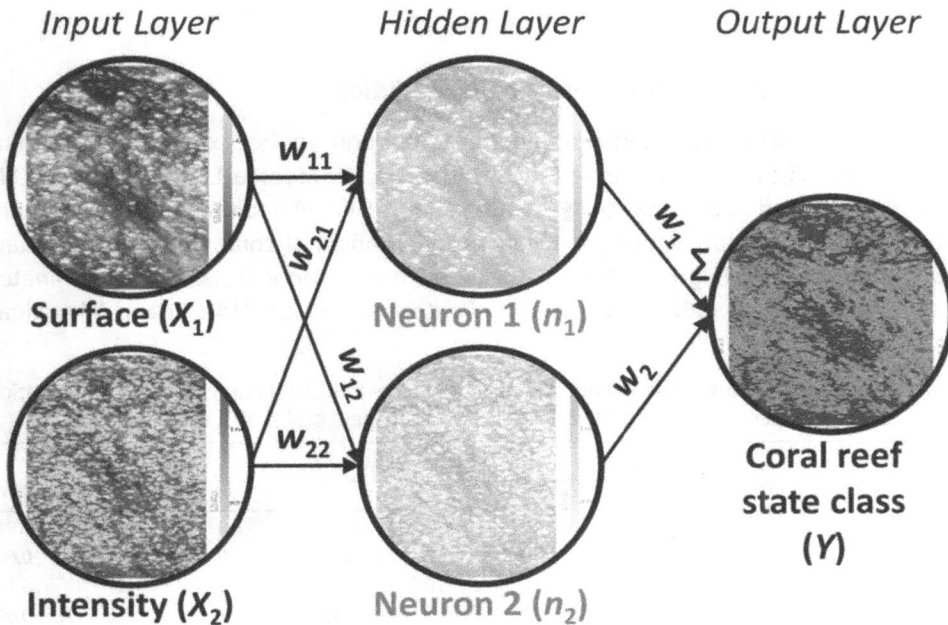

Figure 3. Conceptual flowchart explaining how the combination of LiDAR surface and intensity can predict the ecological state class through an intermediate hidden layer provided with two neurons.

2.5. *Performance analysis*

The agreement between validation and classified pixels in the five ecological states was quantified using the confusion matrix, from which overall, producer's and user's accuracies (OA, PA and UA, respectively) were computed (Congalton and Green 2009). PA and UA were calculated as the correctly classified pixels in each coral state divided by the number of calibration pixels of the corresponding state, and the total number of pixels that were classified in that state, respectively. OA was reckoned as the correctly classified pixels in all states divided by the total number of pixels.

3. Results

3.1. *Local coral reef state at very high resolution*

The OA of the ANN classification reached a satisfactory performance (OA = 0.75), showing that the dual combination of LiDAR surface and intensity variables had a robust explanatory power of the variability of coral reef states (Table 2). Contrary to coral reef states 1, 5 and 3 that were adequately assigned (UA = 0.84, 0.80 and 0.74, respectively), intermediate coral reef states 2 and 4 were moderately classified with UA of 0.68 and 0.68, respectively (Table 2). Contrary to UA statistics, PA measures were evenly correct (From 0.81 to 0.71, Table 2). The ANN classifier was applied to each pixel of LiDAR DSM and DIM (Figure 2(b, c), respectively) in order to continuously map ecological state (Figure 4(b)) provided with 0.5 m spatial resolution (142 × 422 pixels).

3.2. *Moorea coral reef state at very high resolution*

Insofar as the ANN prediction was adequate enough to be extended, the digital ecological classification was mapped at the island scale. Moorea LiDAR DSM and DIM were first rasterized at 0.5 m spatial resolution (Figure 5(a, b)) and then entered as inputs to the ANN classification, which produced a digital model of coral reef ecological state over the whole island (Figure 5(c), 40,364 × 34,588 pixels). Moorea classes are dominated by sand on pavement (56.8%), followed by *Porites* stony corals (14.1%) and Microalgae

Table 2. Confusion matrix synthesizing the quality of the artificial neural network classification applied to the independent 135 validation pixels (27 pixels per coral reef state).

	STATE	Reference class 1	2	3	4	5	Total	UA
Classified class	1	21	2	2	0	0	25	0.84
	2	4	17	3	1	0	25	0.68
	3	1	3	20	2	1	27	0.74
	4	0	1	1	19	7	28	0.68
	5	0	1	2	3	24	30	0.80
	Total	26	24	28	25	32	135	
	PA	0.80	0.71	0.71	0.76	0.75		

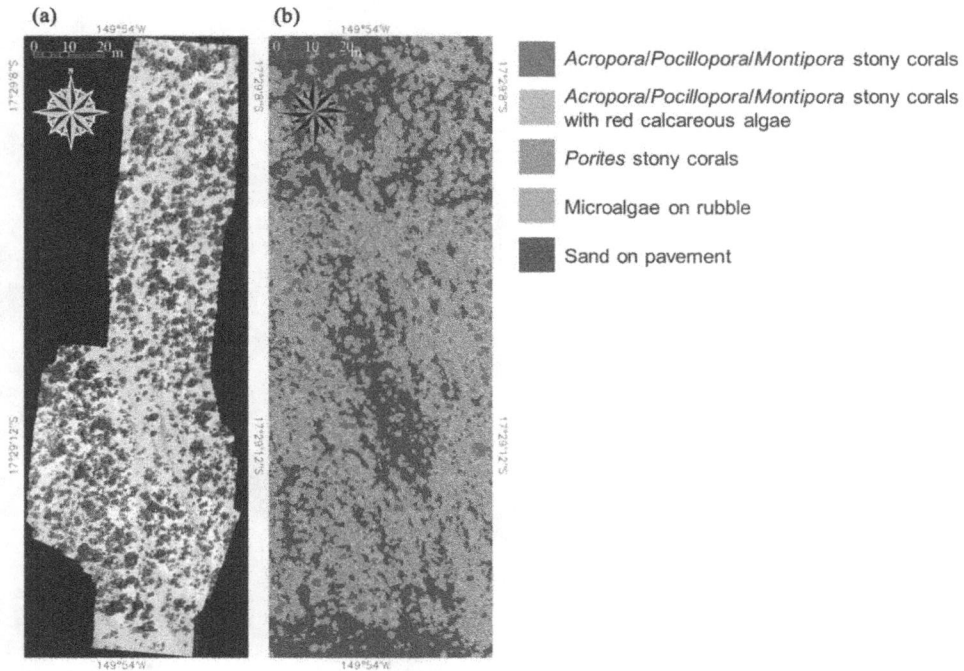

Figure 4. (a) Natural-coloured (blue-green-red) airborne drone orthomosaic (2133 × 6095 pixels, 0.03 m pixel size), along with (b) digital coral reef state classification based on the drone response, LiDAR surface and intensity predictors and two-neuroned artificial neural network classifier (142 × 422 pixels with 0.5 m pixel size).

on rubble (13.8%), then *Acropora/Pocillopora/Montipora* stony corals with red calcareous algae (10.9%), and finally *Acropora/Pocillopora/Montipora* stony corals (4.4%). Overall, the coverage of hard corals (from state 1 to 3) appears significantly greater in the leeward side than the windward side.

4. Discussion

4.1. *Airborne drone as 'air-truth'*

The five coral reef ecological states were based on VHR BGR orthorectified mosaic derived from a consumer-grade multispectral camera driven by an airborne drone. This innovative procedure is supported by our knowledge of in situ coral reef features that can be discriminated at the centimetre scale. Insofar as both ecological composition and structural complexity are easily deduced from the BGR dataset, relatively inexpensive drone deployment can be used to obtain air-truth data directly even in places with little technical capacity. The geolocated photographs can be remotely processed and analysed in the cloud, given a suitable internet connection. With an easy-to-implement flight planning mobile application, rapid surveys could be conducted at even very remote locations with little infrastructure/capacity after short-term events such as cyclone/storm and bleaching. The number of states could be increased by either flying

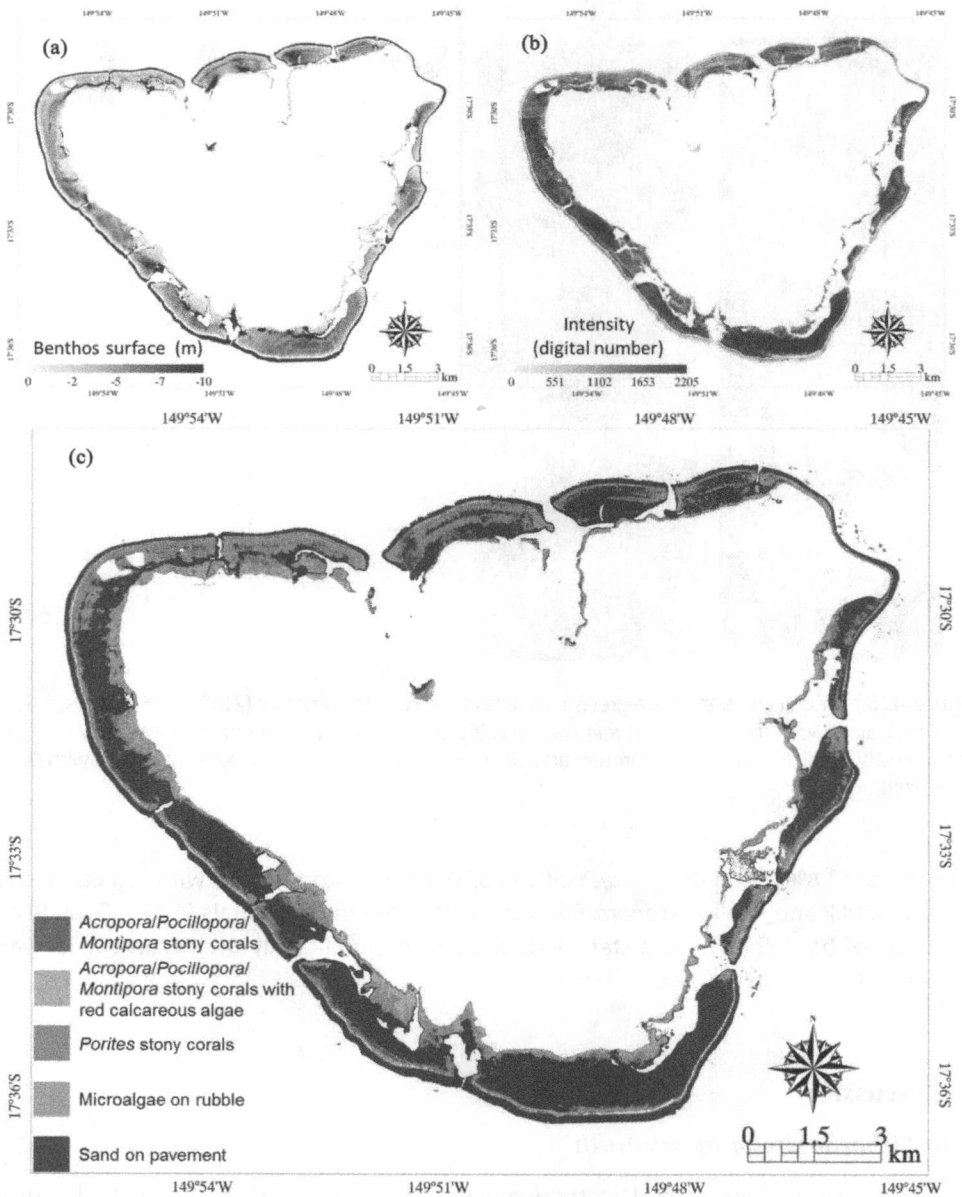

Figure 5. Digital (a) surface, (b) intensity (532 nm wavelength), and (c) coral reef state classification derived from bathymetric LiDAR soundings (40,364 × 34,588 pixels at 0.5 m pixel size).

at lower altitude (to gain in spatial resolution) or using drone-mounted LiDAR that can enhance the vertical accuracy, for example, to differentiate coral from macroalgae (Leiper et al. 2014).

The use of this air-truth, in the form of a cost-efficient UAV-borne BGR orthomosaic, has a strong potential to be applicable to other worldwide coral lagoons and even to a large panel of coastal and aquatic areas, provided with relatively clear waters. This air-truth leverages a high ratio of covered space unit per time unit while collecting centimetre-scale data, considerably outperforming submerged acquisitions, hindered by the very high viscosity of water.

4.2. *Airborne lidar surface and intensity*

The gradient of ecological states (from 1, well-developed hard coral, to 5, sand) was positively correlated with both surface ($r = 0.93$) and intensity ($r = 0.93$), showing that coral coverage decreases with depth and LiDAR green reflectance. The coral shrinkage with depth can be explained by the coral growth and structural complexification towards the surface (as a photosynthetic symbiont), what corroborates results derived from a space-borne reef health proxy (Collin, Hench, and Planes 2012). The negative trend between coral state and green reflectance coincides with in situ spectral measurements, making explicit a greater reflectance of increasingly depigmented blue- and brown-mode coral reefs in the coral health chart (Leiper et al. 2009). This increase in green reflectance (decrease in green absorbance) is linked to the loss of peridinin pigments contained in symbiotic zooxanthellae living in coral tissues (Collin and Planes 2012). Even if most bathymetric LiDAR systems use the single green wavelength, this electromagnetic radiation is relevant to distinguish coral reef state as highlighted in the elaboration of both the green-purple and the 'red edge'-green normalized difference ratios (Collin, Hench, and Planes 2012; Collin, Archambault, and Planes 2014; respectively).

Inner classification results (UA) revealed that coral- and sand-dominant states (1, 3 and 5) were successfully recognized, contrary to both assemblages of corals and rubble colonized by calcareous and micro-algae (2 and 4), respectively. We could assume that the spectral mixing due to the presence of algae on relatively 'pure' states was not very effectively resolved by the ANN classifier built from only LiDAR surface and intensity. We advocate the experiment of a coral reef state classification using an innovative bathymetric LiDAR, augmented by an added spectral wavelength (i.e. 355 nm, as the third harmonic of the 1064-nm laser), likely to detect the coral fluorescence as well as intermediate states (Sasano et al. 2012).

4.3. *Moorea coral reef states' spatial patterns*

The coral reef state classification, spatially classified at VHR, is a strong asset to outline hotspots of healthy coral reefs, thus of associated biodiversity and ecosystem services. The centimetre and decimetre scales targeted in this study greatly enhance the spatial resolution of coral reefs' diagnoses and prognoses, surpassing other recent studies using object-based image analysis, which bottom at 2 m or 10 m (Phinn, Roelfsema, and Mumby 2012; Roelfsema et al. 2013). LiDAR-based spatially explicit classification, provided with decimetre sounding density over 100 km^2, offers an unpublished map of Moorea coral reefs' health. Two main spatial patterns emerged from the spatially explicit classification: westward polarization of healthy fringing reefs and northward polarization of healthy barrier reefs.

Wide healthy fringing reefs along west shorelines strongly contrast with thin ones along the eastern coast. This outstanding geographic difference is very susceptible to be the consequence of the dominant easterly winds (i.e. Southeast trade winds), which entail significantly greater amounts of rain then carried sediment, which, in turn, deposit onto and stress coral colonies (Fabricius 2005), impeding the development of eastern fringing coral reefs.

More extended barrier reefs are obvious in the northern compared to the southern lagoon. This patterning might be explained by the two dominant swell systems originating from South: 40% SE and 25% SSW (Etienne 2012). Swell average height tends to be higher than 4 m during Austral winter, what creates, at the reef, significant wave height greater than 8 m (e.g. Teahupoo spot in Southern Tahiti Iti, Etienne 2012). The exposure to this high to very high energy flow hinders the efficient settlement of coral larvae and breaks the coral assemblage structure (Madin and Connolly 2006). This interpretation is corroborated by the third dominant swell system (22% NE, Etienne 2012), which constrains NE lagoon to exhibit slightly less extended barrier reefs compared to NW.

5. Conclusion

This original research has demonstrated that airborne bathymetric LiDAR data are able to reliably map five ecological states in coral reef systems at VHR over shallow, clear waters. Reef state information can be gleaned from an airborne drone equipped with a multispectral imaging sensor. Novel findings can be summarized as follows:

(1) Coral reef state at the colony-scale (pixel size = 0.03 m) can be sourced from a BGR camera mounted on an airborne low-altitude drone;
(2) LiDAR surface and intensity are powerful predictors of coral reef ecological state at the colony-scale (pixel size = 0.03 m);
(3) ANN is an efficient classification approach to predict ecological state based only LiDAR surface and intensity (OA = 0.75);
(4) LiDAR surface and intensity are powerful predictors of ecological state at the landscape scale (pixel size = 0.5 m);
(5) Healthy fringing and barrier coral reefs in Moorea are located on the western and northern parts of the lagoon, respectively.

Acknowledgments

Authors gratefully thank *Service Hydrographique et Océanographique de la Marine* for the LiDAR acquisition control and the IDEA Consortium for sparking this collaborative research. This work was partly supported by French Polynesia government for LiDAR acquisition, the ETH Zurich for purchasing satellite imagery, and the National Science Foundation through the Moorea Coral Reef LTER (OCE-1236905 and 1637396) and Physical Oceanography (OCE-143133) programs. Two valuable referees and the editor are deeply acknowledged for the manuscript improvement.

Disclosure statement

No potential conflict of interest was reported by the authors.

Funding

This work was supported by the National Science Foundation [OCE-1236905,OCE-143133,OCE-1637396].

References

Adjeroud, M., F. Michonneau, P. J. Edmunds, Y. Chancerelle, T. L. De Loma, L. Penin, L. Thibaut, J. Vidal-Dupiol, B. Salvat, and R. Galzin. 2009. "Recurrent Disturbances, Recovery Trajectories, and Resilience of Coral Assemblages on a South Central Pacific Reef." *Coral Reefs* 28 (3): 775–780. doi:10.1007/s00338-009-0515-7.
Casella, E., A. Collin, D. Harris, S. Ferse, S. Bejarano, V. Parravicini, J. L. Hench, and A. Rovere. 2017. "Mapping Coral Reefs Using Consumer-Grade Drones and Structure from Motion Photogrammetry Techniques." *Coral Reefs* 36 (1): 269–275. doi:10.1007/s00338-016-1522-0.
Collin, A., B. Long, and P. Archambault. 2011. "Benthic Classifications Using Bathymetric LIDAR Waveforms and Integration of Local Spatial Statistics and Textural Features." *Journal of Coastal Research* 62: 86–98. doi:10.2112/SI_62_9.
Collin, A., J. Laporte, B. Koetz, F. R. Martin-Lauzer, and Y. L. Desnos. 2016. "Mapping Bathymetry, Habitat, and Potential Bleaching of Coral Reefs Using Sentinel-2." In *Proceedings of the 13th International Coral Reef Symposium*, Honolulu, 373–387.
Collin, A., J. L. Hench, and S. Planes. 2012. "A Novel Spaceborne Proxy for Mapping Coral Cover." In *Proceedings of the 12th International Coral Reef Symposium*, Cairns, 1–5.
Collin, A., K. Nadaoka, and L. Bernardo. 2015. "Mapping the Socio-Economic and Ecological Resilience of Japanese Coral Reefscapes across a Decade." *ISPRS International Journal of Geo-Information* 4 (2): 900–927. doi:10.3390/ijgi4020900.
Collin, A., N. Lambert, and S. Etienne. 2018. "Satellite-Based Salt Marsh Elevation, Vegetation Height, and Species Composition Mapping Using the Superspectral WorldView-3 Imagery." *International Journal of Remote Sensing* 1–19. doi:10.1080/01431161.2018.1466084.
Collin, A., P. Archambault, and B. Long. 2008. "Mapping the Shallow Water Seabed Habitat with the SHOALS." *IEEE Transactions on Geoscience and Remote Sensing* 46 (10): 2947–2955. doi:10.1109/TGRS.2008.920020.
Collin, A., P. Archambault, and B. Long. 2011. "Predicting Species Diversity of Benthic Communities within Turbid Nearshore Using Full-Waveform Bathymetric LiDAR and Machine Learners." *PloS one* 6 (6): e21265. doi:10.1371/journal.pone.0021265.
Collin, A., P. Archambault, and S. Planes. 2014. "Revealing the Regime of Shallow Coral Reefs at Patch Scale by Continuous Spatial Modeling." *Frontiers in Marine Science* 1: 65. doi:10.3389/fmars.2014.00065.
Collin, A., S. Etienne, and E. Feunteun. 2017. "VHR Coastal Bathymetry Using WorldView-3: Colour versus Learner." *Remote Sensing Letters* 8 (11): 1072–1081. doi:10.1080/2150704X.2017.1354261.
Collin, A., and S. Planes. 2012. "Enhancing Coral Health Detection Using Spectral Diversity Indices from Worldview-2 Imagery and Machine Learners." *Remote Sensing* 4 (10): 3244–3264. doi:10.3390/rs4103244.
Congalton, R. G., and K. Green. 2009. *Assessing the Accuracy of Remotely Sensed Data: Principles and Practices*. Boca Raton, FL: CRC /Taylor Francis press.
Costa, B. M., T. A. Battista, and S. J. Pittman. 2009. "Comparative Evaluation of Airborne LiDAR and Ship-Based Multibeam SoNAR Bathymetry and Intensity for Mapping Coral Reef Ecosystems." *Remote Sensing of Environment* 113 (5): 1082–1100. doi:10.1016/j.rse.2009.01.015.
Cressey, D. 2015. "Tropical Paradise Inspires Virtual Ecology Lab." *Nature* 517: 255–256. doi:10.1038/517255a.
Davies, N., D. Field, D. Gavaghan, S. J. Holbrook, S. Planes, M. Troyer, M. Bonsall, et al. 2016. "Simulating Social-Ecological Systems: The Island Digital Ecosystem Avatars (IDEA) Consortium." *Gigascience* 5: 14. doi:10.1186/s13742-016-0118-5.

Etienne, S. 2012. "Marine Inundation Hazards in French Polynesia: Geomorphic Impacts of Tropical Cyclone Oli in February 2010." *Geological Society, London, Special Publications* 361 (1): 21–39. doi:10.1144/SP361.4.

Fabricius, K. E. 2005. "Effects of Terrestrial Runoff on the Ecology of Corals and Coral Reefs: Review and Synthesis." *Marine Pollution Bulletin* 50 (2): 125–146. doi:10.1016/j.marpolbul.2004.11.028.

Goodman, J. A., J. P. Samuel, and R. P. Stuart. 2013. *Coral Reef Remote Sensing. A Guide for Mapping, Monitoring and Management.* Netherlands: Springer.

Hedley, J. D., C. M. Roelfsema, I. Chollett, A. R. Harborne, S. F. Heron, S. Weeks, W. J. Skirving, et al. 2016. "Remote Sensing of Coral Reefs for Monitoring and Management: A Review." *Remote Sens* 8: 118. doi:10.3390/rs8020118.

Heermann, P. D., and N. Khazenie. 1992. "Classification of Multispectral Remote Sensing Data Using a Back-Propagation Neural Network." *IEEE Transactions on Geoscience and Remote Sensing* 30 (1): 81–88. doi:10.1109/36.124218.

Knudby, A., S. Jupiter, C. Roelfsema, M. Lyons, and S. Phinn. 2013. "Mapping Coral Reef Resilience Indicators Using Field and Remotely Sensed Data." *Remote Sensing* 5 (3): 1311–1334. doi:10.3390/rs5031311.

Leiper, I. A., S. R. Phinn, C. M. Roelfsema, K. E. Joyce, and A. G. Dekker. 2014. "Mapping Coral Reef Benthos, Substrates, and Bathymetry, Using Compact Airborne Spectrographic Imager (CASI) Data." *Remote Sensing* 6 (7): 6423–6445. doi:10.3390/rs6076423.

Leiper, I. A., U. E. Siebeck, N. J. Marshall, and S. R. Phinn. 2009. "Coral Health Monitoring: Linking Coral Colour and Remote Sensing Techniques." *Canadian Journal of Remote Sensing* 35 (3): 276–286. doi:10.5589/m09-016.

Leon, J. X., C. M. Roelfsema, M. I. Saunders, and S. R. Phinn. 2015. "Measuring Coral Reef Terrain Roughness Using 'Structure-From-Motion' Close-Range Photogrammetry." *Geomorphology* 242: 21–28. doi:10.1016/j.geomorph.2015.01.030.

Madin, J. S., and S. R. Connolly. 2006. "Ecological Consequences of Major Hydrodynamic Disturbances on Coral Reefs." *Nature* 444 (7118): 477–480. doi:10.1038/nature05328.

Pastol, Y., L. Chamberlain, and M. Sinclair. 2016. "Airborne Bathymetric LiDAR and Coastal Zone Management in French Polynesia." In *Proc. of International Federation of Surveyors (FIG) Working Week*, Christchurch, May 2–6.

Phinn, S. R., C. M. Roelfsema, and P. J. Mumby. 2012. "Multi-Scale, Object-Based Image Analysis for Mapping Geomorphic and Ecological Zones on Coral Reefs." *International Journal of Remote Sensing* 33 (12): 3768–3797. doi:10.1080/01431161.2011.633122.

Roelfsema, C., S. Phinn, S. Jupiter, J. Comley, and S. Albert. 2013. "Mapping Coral Reefs at Reef to Reef-System Scales, 10s–1000s Km2, Using Object-Based Image Analysis." *International Journal of Remote Sensing* 34 (18): 6367–6388. doi:10.1080/01431161.2013.800660.

Rowlands, G., S. Purkis, B. Riegl, L. Metsamaa, A. Bruckner, and P. Renaud. 2012. "Satellite Imaging Coral Reef Resilience at Regional Scale. A Case-Study from Saudi Arabia." *Marine Pollution Bulletin* 64 (6): 1222–1237. doi:10.1016/j.marpolbul.2012.03.003.

Sasano, M., H. Yamanouchi, A. Matsumoto, N. Kiriya, K. Hitomi, and K. Tamura. 2012. "Development of Boat-Based Fluorescence Imaging Lidar for Coral Monitoring." In *Proc. of 12th Internaitonal Coral Reef Symposium*, Cairns, 5A–7.

Service Hydrographique et Océanographique de la Marine (SHOM). 2016. *Lidar Polynésie française 2015 Produit Moorea SAU V. 20160630.* Brest, France: SHOM press.

Wedding, L. M., A. M. Friedlander, M. McGranaghan, R. S. Yost, and M. E. Monaco. 2008. "Using Bathymetric Lidar to Define Nearshore Benthic Habitat Complexity: Implications for Management of Reef Fish Assemblages in Hawaii." *Remote Sensing of Environment* 112 (11): 4159–4165. doi:10.1016/j.rse.2008.01.025.

A comparison of airborne hyperspectral-based classifications of emergent wetland vegetation at Lake Balaton, Hungary

Dimitris Stratoulias ⓘ, Heiko Balzter, András Zlinszky and Viktor R. Tóth

ABSTRACT

Earth observation has rapidly evolved into a state-of-the-art technology providing new capabilities and a wide variety of sensors; nevertheless, it is still a challenge for practitioners external to a specialized community of experts to select the appropriate sensor, define the imaging mode requirements, and select the optimal classifier or retrieval method for the task at hand. Especially in wetland mapping, studies have relied largely on vegetation indices and hyperspectral data to capture vegetation attributes. In this study, we investigate the capabilities of a concurrently acquired very high spatial resolution airborne hyperspectral and lidar data set at the peak of aquatic vegetation growth in a nature reserve at Lake Balaton, Hungary. The aim was to examine to what degree the different remote-sensing information sources (i.e. visible and near-infrared hyperspectral, vegetation indices and lidar) are contributing to an accurate aquatic vegetation map. The results indicate that de-noised hyperspectral information in the visible and very near-infrared bands (400–1000 nm) is performing most accurately. Inclusion of lidar information, hyperspectral infrared bands (1000–2500 nm), or extracted vegetation indices does not improve the classification accuracy. Experimental results with algorithmic comparisons show that in most cases, the Support Vector Machine classifier provides a better accuracy than the Maximum Likelihood.

1. Introduction

In situ data are the most reliable source of information for aquatic species vegetation mapping; however, a proximate field collection scheme for a large geographic area at frequent intervals is a cumbersome task, if not impossible. Earth observation, synergistically with field measurements, is the only efficient way of monitoring a large wetland or land areas covered by vegetation patches. However, the high biodiversity encountered in the ecotone between terrestrial and aquatic ecosystems can result in a complex spatial structure and a lack of crisp boundaries between habitat types. This high variability of plant species and their spatial distribution requires information acquired at a fine scale and with a high radiometric discriminatory capability. Such a case is the aquatic vegetation around Lake Balaton, Hungary, which is the largest (596 km^2) freshwater lake in Central Europe (Virág 1997). It encompasses a total area of approximately 11 km^2 of reeds stretching along 112 km of the shoreline and has suffered intense reed dieback from 1970s onwards (Kovács et al. 1989).

Wetland mapping has been an application domain for remote sensing from the late 1960s and onwards with the means of aerial photographs and satellite images since the advent of the Landsat satellites constellation (Bartlett and Klemas 1980; Butera 1983). The need to establish advanced methodologies for wetland vegetation mapping based on remote-sensing data has been frequently stressed (e.g. Rebelo, Finlayson, and Nagabhatla 2009) and there are indications that species detection and patterns of species richness are benefited greatly from remote-sensing data (Turner et al. 2003). Medium-resolution multispectral Landsat satellite data have been used in this context (e.g. Baker et al. 2006; Reschke and Hüttich 2014; Robertson, King, and Davies 2015; Dvorett, Davis, and Papeş 2016) and lately Sentinel-2 data have complemented and augmented the Earth observation tools for wetland mapping (e.g. Stratoulias et al. 2015; Kaplan and Avdan 2017; Pereira, Melfi, and Montes 2017). For an overview of wetland mapping, Adam, Mutanga, and Rugege (2010) studied the identification of wetland vegetation review of hyperspectral data on wetlands, Ozesmi and Bauer (2002) discussed the classification schemes used in remote sensing of wetlands, and Klemas (2011) compiled a comprehensive review of practical techniques.

Hyperspectral remote sensing is pivotal for vegetation mapping due to its ability to discriminate vegetation types based on their narrowband spectral characteristics, the latter correlating with pigments and cannot easily be differentiated with multispectral sensors. Currently only a few hyperspectral spaceborne imagers are orbiting around the Earth, namely the sensor Compact High Resolution Imaging Spectrometer on board Project for On-Board Autonomy-1, Hyperion on board Earth-Observing One, and hyperspectral imagery on board China Environment 1A series, but more will become available in the near future, namely the Environmental Mapping and Analysis Program (Guanter et al. 2015), the Deutsches Zentrum für Luft- und Raumfahrt Earth Sensing Imaging Spectrometer (Eckardt et al. 2015), the PRecursore IperSpettrale della Missione Applicativa (Stefano et al. 2013), the Spaceborne Hyperspectral Applicative Land and Ocean Mission (Feingersh and Dor 2015), the Hyperspectral Infrared Imager (Lee et al. 2015), and the Hyperspectral Imager Suite (Kashimura et al. 2013). However, the trade-off of the spectral fidelity at the expense of the spatial resolution deem such data sets unsuitable for studies focusing on small areas, hence the deployment of airborne hyperspectral sensors is necessary. For instance, Airborne Visible/ Infrared Imaging Spectrometer data have been proven to be advantageous in areas covered by vegetation with similar spectral characteristics, such as wetlands (Neuenschwander, Crawford, and Provancha 1998). Hirano, Madden, and Welch (2003) used the same data source to create a vegetation map of a wetland park in Florida, USA, reporting success in identifying vegetation communities as well as detecting invasive exotic species. Malthus and George (1997) suggested that airborne remote sensing has a strong potential for monitoring freshwater macrophyte species. In this context, Burai et al. (2010) separated seven vegetation classes in a wetland in Hungary with overall accuracy 78% and κ = 0.63. In the same paper, they stress the need to develop wetland-specific spectral libraries. Last but not least, in a study investigating aquatic reed, Gilmore et al. (2009) reported high classification accuracies due to unique high near-infrared reflectance in early autumn.

Along hyperspectral reflectance, the plant height can be a distinguishing factor between homogeneous patches of macrophytic vegetation, and between macrophytes and other vegetated/non-vegetated objects in the scene. Lidar can provide information about the height and structure of the canopy, which is an independent and complementary source of information to spectroscopy. Lidar has been used in forestry due to the ability to penetrate into the canopy, allowing estimation of height and volumetric distribution of trees (e.g. Miller 2001; Bradbury et al. 2005; Hinsley et al. 2006; Balzter et al. 2007; Puttonen et al. 2010; Jones, Coops, and Sharma 2010; Pedergnana et al. 2011). However, despite the fact that tree canopies are composed of plants which vary in species, height, size, and texture, macrophytes are encountered most often in assemblages of species with more homogeneous spatial characteristics. Anderson et al. (2010) have stressed the usefulness of lidar data in mapping habitat architecture in a range of ecosystems by capturing fine spatial patterns. Downward-looking lidar from airborne platforms has recently been used in the context of wetlands for discriminating vegetation species from marsh components (Rosso, Ustin, and Hastings 2006), classifying wetland elements based solely on lidar data (Brennan and Webster 2006), estimating the inundation level below forest canopy based on the amplitude of the lidar signal (Lang and McCarty 2009), synergistically with high-resolution satellite data to improve wetland distinction (Maxa and Bolstad 2009) and categorizing vegetation and alkali grassland associations in Lake Balaton, Hungary (Zlinszky et al. 2014).

With satellite data available at increasingly high spatial resolutions and decreasing cost, recent research has focused on fusing optical satellite images with airborne lidar data in order to increase the classification accuracy. The underlying idea is that data acquired from different sensors provide information on different properties of the vegetation (for instance, hyperspectral provide information on the biochemistry and lidar on the structure of canopy properties (Koetz et al. 2007)), which to an extent are independent from and complementary to each other. Dalponte, Bruzzone, and Gianelle (2008) presented a methodology for fusing hyperspectral and lidar data in vegetation studies and urged to develop advanced classification systems integrating these two data sources. A few studies have used hyperspectral and lidar data sets synergistically (e.g. Hunter et al. 2010; Onojeghuo and Blackburn 2011; Mewes, Franke, and Menz 2011; Lausch et al. 2013). Several authors have reported increased classification accuracy as a result of data integration from multiple sensors. Jones, Coops, and Sharma (2010) mapped 11 tree species in a coastal region and report an increase in users' accuracy by 8.4–18.8% after data fusion in comparison to using solely hyperspectral data. Johansen et al. (2010) presented an automatic feature extraction of biophysical properties from lidar data and suggest that similar applications can be employed by natural resource management agencies. Swatantran et al. (2011) concluded that these two data types have many potential applications in ecological and habitat studies. Klemas (2011) recommended that the combined use of lidar and hyperspectral imagery can improve the accuracy of wetland species discrimination. Onojeghuo and Blackburn (2011) optimized the synergistic use of lidar and AISA hyperspectral data for mapping reed bed habitats and report a significant accuracy improvement by 11% when a lidar-derived mask is incorporated. Finally, Niculescu et al. (2016) reported an increase in mean accuracy of 14% when combining lidar with Satellite Pour l'Observation de la Terre or Earth-observing Satellites (SPOT) multispectral satellite images compared to using only SPOT images.

The increasing availability of Earth observation data has resulted in rapidly growing data archives; however, more data does not necessarily translate into better analysis (Fernandez-Prieto et al. 2006) unless the methods of turning data to information are optimized. Appropriate image processing is an essential aspect of any classification of remotely sensed data to a meaningful categorical map (Lu and Weng 2007). Pixel-based classifiers have traditionally been the means to classify remote-sensing imagery. These classifiers are making categorical judgements based solely on the spectral information of each individual pixel (Gong and Howarth 1990). The most widely used supervised classifier in remote sensing has historically been the Maximum Likelihood (ML) classifier which assumes Gaussian distributions of the spectral reflectance values within each class and for each spectral band.

Another supervised non-parametric statistical learning technique that has gained popularity over the last years (e.g. Brown, Lewis, and Gunn 2000; Keramitsoglou et al. 2006; Chi, Feng, and Bruzzone 2008) is the Support Vector Machine (SVM), developed originally by Vapnik (1995). SVM performs well in cases with a small number of training samples, which is a frequent problem in remote-sensing classifications. Several authors have confirmed this superiority of SVM over alternative classifiers when applied to hyperspectral data (Melgani and Bruzzone 2004; Pal and Mather 2004; Camps-Valls and Bruzzone 2005; Oommen et al. 2008; Dalponte, Bruzzone, and Gianelle 2008; Hunter et al. 2010; Bahria, Essoussi, and Limam 2011; for a review the reader is directed to Mountrakis, Im, and Ogole (2011)) and for land-use/land-cover classifications (Pal and Mather 2005; Boyd, Sanchez-Hernandez, and Foody 2006; Keramitsoglou et al. 2006; Dixon and Candade 2008; Dalponte et al. 2009). In the framework of reed detection, Yang et al. (2012) compared six classifiers for airborne hyperspectral imagery for mapping giant reed and confirmed the suitability of SVM. Furthermore, SVM seems to eliminate the Hughes effect (the predictive power reduces as the spectral dimensionality increases, see Hughes 1968), which is crucial for high-dimensional hyperspectral data (Pal and Mather 2004; Oommen et al. 2008) as, for instance, it has been shown to affect other classifiers, such as ML, when dealing with a large number of bands (Pandey, Tate, and Balzter 2014).

The above review of the scientific literature presents the challenges of mapping macrophytes based on remote sensing arising mainly from the fine spatial scale and radiometric accuracy required to unveil the complex structure of the ecosystem and the vegetation cover at species level. A variety of data sets, fusion of data sets, classification algorithms, and methods have been proposed to increase the mapping accuracy. As a consequence, it remains a challenge to investigate the means with which information at high spatial resolution can be translated to accurate information about aquatic vegetation. This study aims to devise a suitable classification scheme for lakeshore vegetation mapping with high spatial resolution airborne data and to quantify its accuracy.

It presents an evaluation of different remotely sensed data sets for classifying lakeshore vegetation classes using two classification algorithms that are common in wetland mapping, namely ML and SVM.

2. Study area

The study area is located at the Bozsai Bay on the northwest part of the Tihany Peninsula of Lake Balaton, Hungary (Figure 1) (Somlyody, Herodek, and Fischer 1983). It encompasses a nature reserve of the Balaton Uplands National Park and as such is relative to

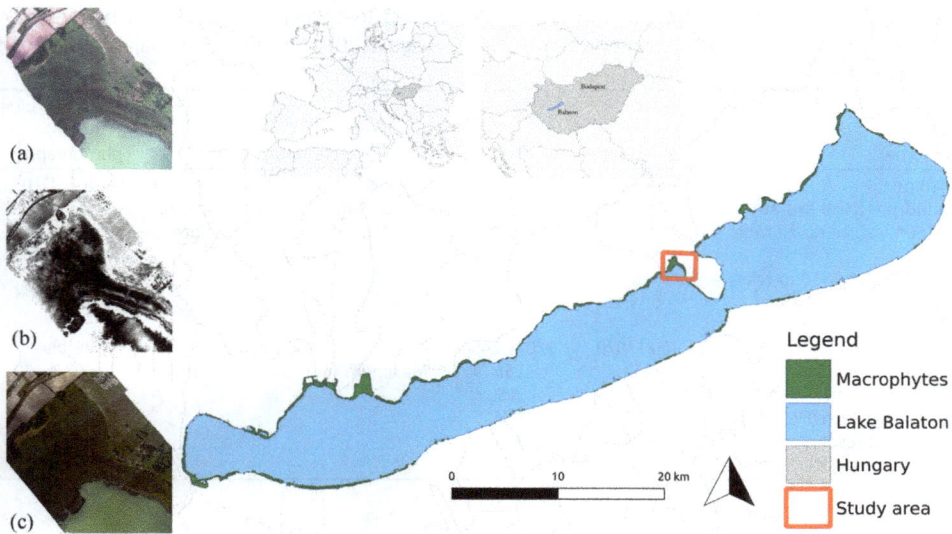

Figure 1. Study area in the Bozsai Bay, Hungary (latitude 46.917899, longitude 17.835806). Inset numbered images depict thumbnails of the concurrently collected (a) hyperspectral image, (b) lidar data, and (c) orthophoto.

other reed beds of Lake Balaton, quasi-undisturbed from human activity. A variety of macrophytes, trees, and grasslands are encountered; however, reed is encountered frequently, especially *Phragmites australis* (Cav.) Trin. ex Steud, and to a smaller extent *Typha angustifolia* L., *Typha latifolia* L. and *Carex* sp.

In the last decades the stability of the waterward fringe of the reed bed has been deteriorating and the whole ecosystem has been retreating from the deep water, a phenomenon known as the 'reed dieback' (Van Der Putten 1997; Stratoulias et al. 2015). As a consequence, the reed bed contains not only a diversity of vegetation species but also vegetation patches of different stability. The study area is situated in the vicinity of the Balaton Limnological Institute, providing easy access for fieldwork and assuring familiarity of the authors and experienced researchers with the local vegetation.

3. Airborne data

A European Facility for Airborne Research (EUFAR) airborne campaign was undertaken from 21 to 26 August 2010 by the NERC Airborne Research Facility by the UK Natural Environment Research Council (NERC). The platform used was a Dornier 228-101 research aircraft which flew at 1550 m above mean sea level and was equipped with an Inertial Measuring Unit/Global Navigation Satellite System providing information on the aircrafts' position and orientation, respectively. The survey covered the whole lakeshore around Lake Balaton and the Kis Balaton, an adjacent wetland to the southwest of the lake. The data set comprised concurrently recorded hyperspectral imagery (400–2500 nm), discrete return lidar data, and orthophotos (Zlinszky et al. 2011). In this study we use data from two adjacent flight strips (Figure 1) acquired on 21 August 2010 between 13:40 and 14:18 GMT (Table 1) over the Bozsai Bay which observes the Central European Summer Time.

Table 1. Specifications of remote-sensing instruments used to collect simultaneously hyperspectral imagery, lidar data, and orthophotos during 21–26 August 2010 around Lake Balaton, Hungary, under clear sky conditions.

Data type	Hyperspectral	Hyperspectral	Discrete return lidar	Red, green, blue (RGB) photography
Instrument	AISA Eagle	AISA Hawk	Leica ALS50-11	Leica RCD105
Ground pixel size (m) at 1550 m absolute altitude	1.50	2.10	4 returns maximum (resampled), 1 pt m^{-2} stripwise point density	0.175
Swathn (m) at 1550 m relative altitude (1445 m true altitude)	992 (38° field of view)	614 (24° field of view)	–	948
Spectral domain (nm)	400 – 970	970 – 2450	1064	Visible
Number of bands	253	256	Maximum four discrete returns	3
Band width (nm)	3.3	8.5	–	RGB
Spectral resolution (FWHM) (nm)	2.20–2.44	6.31	–	–
Radiometric resolution	12 bit	14 bit	–	16 bit
Signal-to-noise ratio (SNR)	1250:1 (max)	800:1 (max)	–	–

The hyperspectral data set was collected from an airplane-mounted Specim AISA dual system (Spectral Imaging Ltd., Oulu, Finland) integrating the nadir-looking sensors Eagle and Hawk as in similar studies (Artigas and Yang 2005; Jensen et al. 2007; Dalponte, Bruzzone, and Gianelle 2008; Shafri and Hamdan, 2009; Yang and Artigas 2009; Burai et al. 2010; Onojeghuo and Blackburn 2011). The two sensors record incoming radiation cumulatively in 509 bands from 400 to 2450 nm with a spectral resolution (in full-width half-maximum (FWHM)) 2.20–2.44 nm and 6.32 nm for Eagle and Hawk, respectively, and delivered a spatial resolution of 1.50 and 2.10 m, respectively (Table 1). The instruments are nadir-looking, and therefore, the angle between the line of sight of the sensors and the zenith is 180°.

Lidar data were recorded from a Leica ALS50 compact laser scanning system. A maximum of four discrete returns at 83 kHz (1064 nm) was delivered. At the last calibration before the 2010 campaign, lidar data were judged against ground control points and a mean error of 3.1 cm and standard deviation of 2.2 cm at an altitude of 1350 m was estimated.

True-colour orthophotos were recorded from a 39 megapixel Leica RCD105 medium format digital camera. The charge-coupled device instrument of the camera recorded radiation in three channels in the visible domain and delivered images in 16-bit TIFF format, with approximate ground resolution of 17.5 cm from 1550 m aircraft true altitude. Geometric registration on a projected coordinate system was implemented by the Technical University of Vienna. This data set was visually inspected synergistically with expert knowledge for selecting training and validation data sets from the hyperspectral images in the processes of classification and accuracy assessment, respectively. Further information on the airborne campaign and the sensors' specifications can be found at Zlinszky et al. (2012).

4. Methodology

4.1. *Preprocessing*

The hyperspectral and lidar data were preprocessed to derive meaningful information associated with lakeshore vegetation before the classification as illustrated in Figure 2.

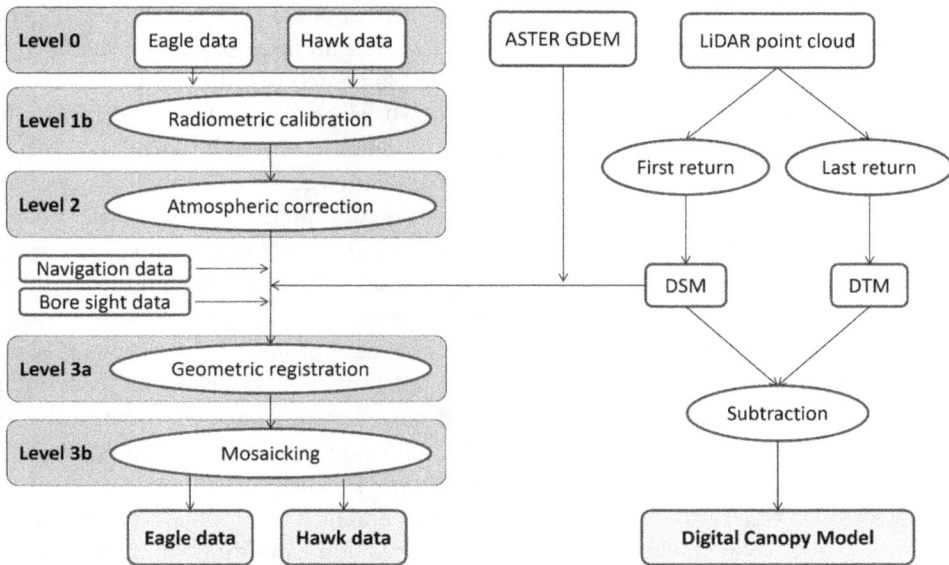

Figure 2. Preprocessing flow chart of the hyperspectral and lidar data sets.

The radiance values of the hyperspectral data were corrected for atmospheric effects, georeferenced, and mosaicked to compose a very fine spectral and spatial resolution image. A digital canopy model (DCM) was derived from the lidar data by subtracting the digital terrain model from the digital surface model (DSM), which are represented by the sensor's last and the first return, respectively. Thereafter the DCM of the two adjacent stripes were mosaicked and resampled to 1.5 m pixel size with the nearest-neighbour interpolation method.

Focusing on the hyperspectral preprocessing, first a radiometric correction was applied to compensate for the effects of the off-nadir surface reflection and glint. Vignetting effects, instrument scanning, off-nadir view angle, and sun reflection as well as other illumination effects can affect the image non-uniformly and are generally regarded as cross-track illumination effects (ITT Visual Information Solutions 2009). In the used data set, the main contributor has been the sun reflection when solar position was diverging from the aircraft orientation (Table 2) and illumination conditions have there-fore not been isotropic. As a result, a glitter at the edge of the images is apparent in flights with north–south orientation. Moreover, the bands for Eagle were restricted to the bandwidth 450–900 nm and for Hawk to 1000–2400 nm as if the quality of the data degrades at the boundaries of the recording spectrum. Along-track mean values were then calculated and plotted to stress the variation in illumination differences across the

Table 2. Solar illumination conditions at Lake Balaton (latitude: 46.9127, longitude 17.8369) during time acquisition as calculated from the National Oceanic and Atmospheric Administration solar calculator (http://www.esrl.noaa.gov/gmd/grad/solcalc).

Flight number (Julian day)	Date	Time (GMT)*	Solar noon	Solar azimuth (°)	Solar elevation (°)
233	21 August 2010	13:40–14:18	12:52	200.16–214.55	53:71–50:71

*Local time is + 1 h from the GMT and observes the Daylight Saving Time.

image lines. Cross-track illumination correction was applied on the samples of each image with a second-order polynomial function and an additive correction method.

Subsequently, atmospheric correction was implemented in the Fast Line-of-sight Atmospheric Analysis of Spectral Hypercubes (FLAASH), which is part of the Environment for Visualizing Images (ENVI) atmospheric correction module (ITT Visual Information Solutions 2009). FLAASH is a first-principles atmospheric correction tool based on a modified version of the MODTRAN4 radiation transfer code (Matthew et al. 2000) applicable in the visible through near-infrared and shortwave infrared regions up to 3 μm. The FLAASH algorithm was selected (Supplemental material) based on the advantage of accounting for the adjacency effect, meaning the occurrence of optical path interference between reflectances from adjacent surface materials (Burazerović et al. 2013). This is especially prominent in coastlines, where existence of water and land, two materials with different spectral behaviour, is merging. Moreover, the phenomenon is stronger at short wavelengths in scenes containing large reflectance contrast (Richter et al. 2006), which is the case for the macrophytes growing in Lake Balaton, the water of which is characterized by high reflectivity in the optical domain due to the suspended sediment, submerged macrophytes, and high chlorophyll concentration of the lake water. Several studies have attempted to develop algorithms for correcting this phenomenon (i.e. Santer and Schmechtig 2000; Sanders, Schott, and Raqueño 2001; Sterckx, Knaeps, and Ruddick 2011).

Considering the time period of the airborne campaign and the geographical position of the study area, the mid-latitude summer atmospheric model was used. No aerosol model was employed, as in the scene there was a lack of dark pixels, which is necessary for the implementation of aerosol integration (Kaufman et al. 1997). Instead, the aerosol amount was estimated from the visibility which was set to 50 km in agreement with the atmospheric conditions on the day of the acquisition and confirmed by the local METAR report. The CO_2 mixing ratio was set to 404 ppm. Spectral polishing was used with a width of nine spectral channels. The ground elevation at the lake is approximately 100 m above sea level and can be assumed constant for the purpose of atmospheric correction as the terrain around the lake is quite flat. The rest of the input for FLAASH was taken from the navigation file recorded during the flight campaign for each individual scene. Overall the atmospheric correction provided typical spectral responses for the vegetation contained in the image (Supplemental material). No artefacts were apparent due to the clear sky conditions at the time of image acquisition and the lack of absorbing bodies that qualify as 'open water' in the spectral sense around the mesotrophic and sediment-laden Bozsai Bay.

Last, geometric registration was applied with the open-source Airborne Processing Library (APL) software v3.1.4 (Warren et al. 2014). The algorithm is designed to geocode the raw imagery by taking into account bore-sight information recorded during the flight. The DSM extracted from the concurrently acquired lidar data set with missing values filled-in from the Advanced Spaceborne Thermal Emission and Reflection Radiometer Global DEM (NASA LP DAAC 2001) was used to increase the geometric precision. The mean absolute along-track and across-track error (at nadir and measured in pixel size) reported from APL developers is 0.39 ± 0.31 and 0.65 ± 0.42, respectively (Warren et al. 2014), which translates for the data set used in this study to 0.58 ± 0.46 m (along track) and 0.97 ± 0.63 m (across track) for Eagle and 0.82 ± 0.65 m (along track) and 1.36 ± 0.88 m (across track) for Hawk. The two adjacent hyperspectral images were registered at the UTM projected coordinate system (Zone 33N) of the World Geodetic System 1984 Datum on a 1.5 m × 1.5 m grid (Figure 3).

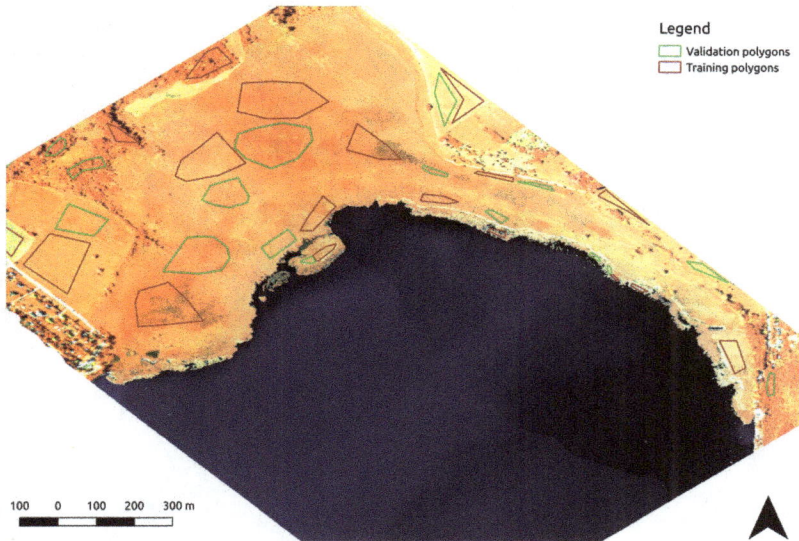

Figure 3. Near IR composite (RGB: 700, 545, and 341 nm) of the hyperspectral Eagle post-processed image and the training and validation polygons used for the first classification of the image in the main vegetation classes.

4.2. *Training and validation sets*

Chen and Stow (2002) made a thorough comparison of common training strategies for classification methods used in the literature and reckon that the training approach can affect the classification result. Moreover, they claim that the training has a higher influence on the result when applied on fine rather than coarse resolution images. Information on vegetation species and geolocation of pure areas formed the basis for selecting the training and validation data sets. For the reed-specific vegetation types, ecologists who are familiar with the study area advised on the definition of the classes. A set of polygons was selected from the hyperspectral image representing the seven emergent-vegetation classes of interest and was verified for homogeneity and representation with the orthophotos. A stratified random sampling (probabilistic method of sampling) was followed for the reason of minimising variability within different zones of the image. The sampling area was divided in large zones and each one was assigned a number of sample units. The position of the units then was defined randomly and the size of the samples was proportional to the size of the class they represent. The concurrently acquired aerial photography was used to assure coherency of the polygons. The set was divided into two groups; the first group comprising 18 polygons (total area 147,228 m^2) was used for training the classifier and the second group comprising 16 polygons (total area 92,920 m^2) for validating the results (Figure 3).

4.3. *Input layers*

The input layers for the classification process were prepared from the Eagle, Hawk, and lidar data (Figure 4). The first input layer was extracted from the Eagle data set while the

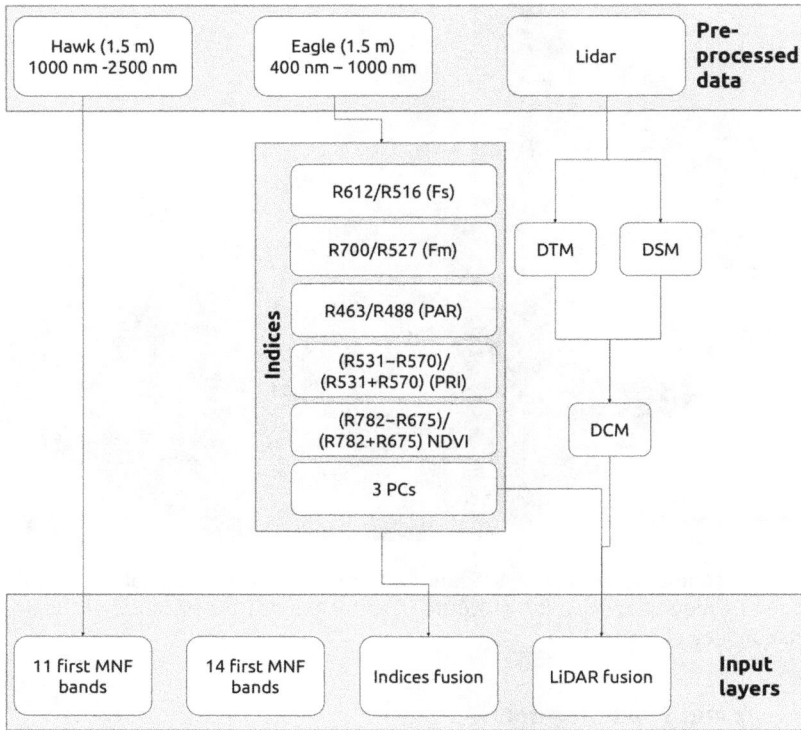

Figure 4. Workflow indicating the preparation of input layers for the classification process based on the Eagle, Hawk, and lidar preprocessed data sets.

second input layer was created in a similar manner from the Hawk data. The third image input for the classification is a layer stack including vegetation indices demonstrating the highest degree of association with reed stability and photosynthetic performance in Lake Balaton as studied by Stratoulias et al. (2015) and popular empirical narrowband indices. These are respectively three reflectance ratios associating to Fs, Fm, and PAR, the photochemical reflectance index (PRI), the normalized difference vegetation index (NDVI), and the first three principal components (PCs) (Figure 4). Finally, the last input layer is based on the three PCs of the Eagle image and the DCM extracted from the lidar data. Two adjacent hyperspectral scenes were used which comprise the reed bed and neighbouring grasslands, agricultural fields (rape, barley, and wheat), roads, settlements, and trees.

The first input layer is solely based on the visible spectrum provided by the Eagle data. Due to the nature of hyperspectral data, the spectral bands are highly correlated and the data set in whole contains a large degree of redundancy. A minimum noise fraction (MNF) transformation (Green et al. 1988) can be applied to eliminate the noise, reduce the dimensionality of the data, and hence the computational requirements without important loss of information. Subsequently, the components (i.e. eigenvalues) of the transformation which are unaffected from noise can be inversed back to the real hyperspace. Pandey, Tate, and Balzter (2014) found that the classification accuracy improves in case ML algorithm is applied on the MNF eigenvalues instead of the total number of bands.

Forward MNF transformation was applied on the Eagle image. Based on the eigenvalues and the corresponding MNF bands, the first 14 transformed bands were selected as the threshold where information is still more prominent in the image in comparison to noise (Figure 5).

The Hawk data are paying in the near-infrared domain and have been acquired simultaneously with the Eagle data; nevertheless, the coverage of the whole reed bed is not complete due to its narrower swath, and adjacent images do not overlap. This appears as a wide missing stripe at the edge of the individual images. Furthermore, Hawk suffers from regular dropped frames, resulting in missing lines in the image. The integration of the Hawk data in the methodology was attempted to evaluate the usefulness of the infrared spectral domain in wetland vegetation mapping. A similar methodology as for Eagle data was followed. Bands between 1336–1462 nm and 1791–1967 nm have been excluded as they are largely affected by atmospheric water vapour absorption. The MNF was then calculated and the first 11 eigenvalues selected in a similar procedure as described in the previous part (Figure 6).

The third data set was based on the assumption that complementary data from lidar and hyperspectral sensors can provide a robust information set. Despite the fact that the DCM derived from moderate-density lidar data underestimates the canopy height (Zlinszky et al. 2012), it is however associated to the canopy height characteristics. The first three PCs from the Eagle image were extracted and combined with the DCM to enhance the information content. The last data set is based on narrowband empirical indices which are associated with reed physiological stability at the specific area of study. The first three PCs from the Eagle image

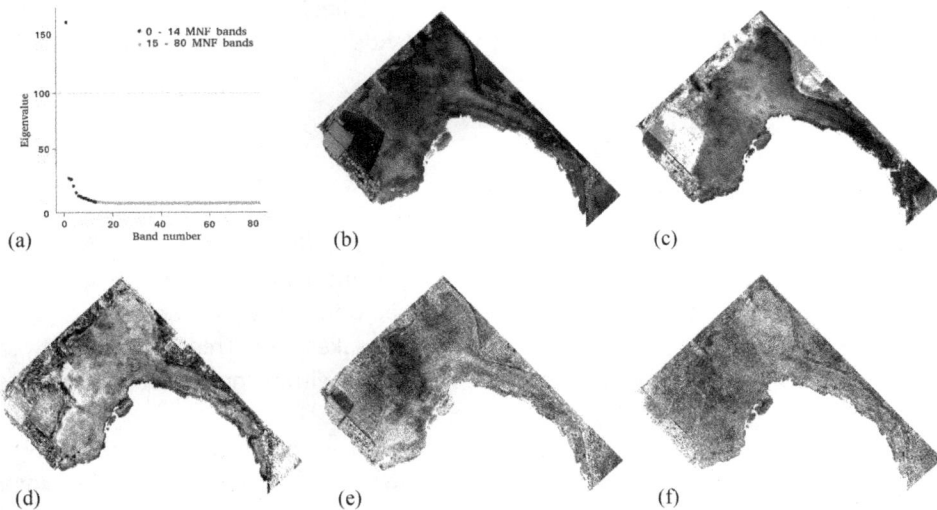

Figure 5. Eigenvalues of the Eagle image (a) and depictions of the minimum noise fraction (MNF) transformation corresponding to band numbers 1 (b), 4 (c), 9 (d), 14 (e), and 16 (f). Noise is considerably higher than the information content after band 14 and hence all the bands after this threshold have been dropped.

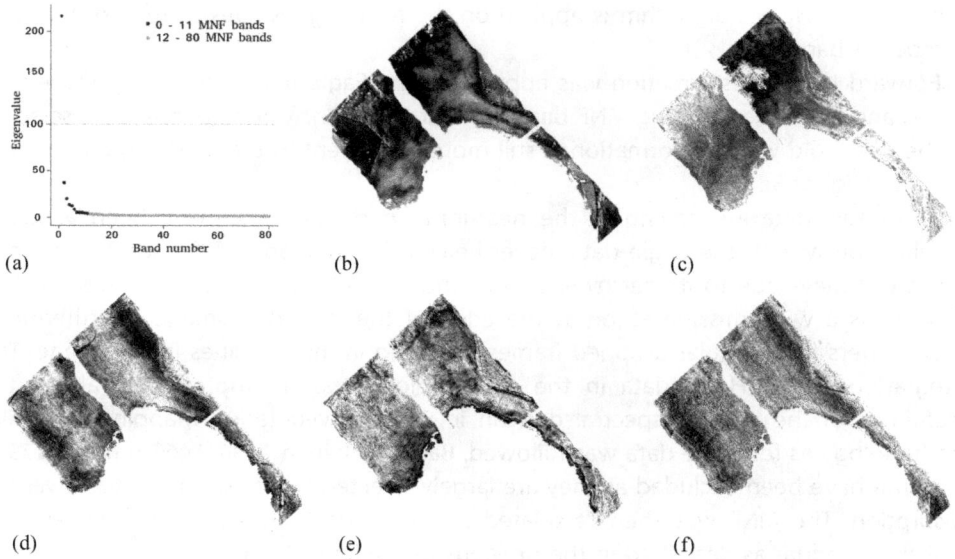

Figure 6. Eigenvalues of the Hawk image (a) and depictions of the minimum noise fraction (MNF) transformation corresponding to band numbers 1 (b), 4 (c), 9 (d), 11 (e), and 12 (f). Noise is considerably higher than the information content after band 11 and hence all the bands after this threshold have been dropped.

were calculated. The narrowband empirical indices NDVI (Tucker 1979) and PRI (Gamon, Penuelas, and Field 1992) were extracted. These are indices heavily used in vegetation-related studies as they are associated to vegetation characteristics. Furthermore, the band rations representing the Fs, Fm′, and PAR per findings of the fluorescence analysis of macrophytes in Lake Balaton as suggested by Stratoulias et al. (2015) were used. The individual layers were combined in a composite image.

4.4. Reed bed masking and mosaicking

The interest of this study lay in the reed bed of the Bozsai Bay. The macrophytes encountered in this nature reserve inherit a diverse and complex structure, and it was decided to narrow the focus on the emergent macrophytes of this area rather than a broader categorization of vegetation types on the lakeshore. Trees, lake water, bare ground, and man-made materials, some of which are within the nature reserve, were excluded.

A mask from the DCM was derived by selecting all pixels with values between 0.3 and 3 m, range which represents typical macrophytic vegetation. The hyperspectral image was subset with the mask to isolate pixels of macrophytes. Finally, a mosaicking procedure was undertaken to stitch together the two images. No colour balancing was used and a feathering distance of 100 pixels was assumed.

4.5. Classification

Several processing steps were iteratively tested and led to the consolidated methodo-logical workflow presented in Figure 7. The classification procedure was conducted identically for the four different input layers developed and using the same training data set. The image was first classified based on the dominant macrophytes encoun-tered in the area. These classes were *Phragmites*, *Typha*, *Carex*, and grassland (this latter constituted by herbaceous (non-woody)) vegetation managed by summer mowing at least once a year. The *Phragmites* class was used to subset the original input layer again, and this subset image was then classified based on the dominance of *Phragmites* in the patch according to the classes of dominant, co-dominant, subdominant, and reed die-back. Reed dominant is the category in which *Phragmites* represents at least 80% of the plant cover density as observed during fieldwork. Reed co-dominant is the category with at least 50% of *Phragmites*. When other macrophytes such as *Typha* and *Carex* are dominant and there is a minor coverage of *Phragmites*, then this defines the

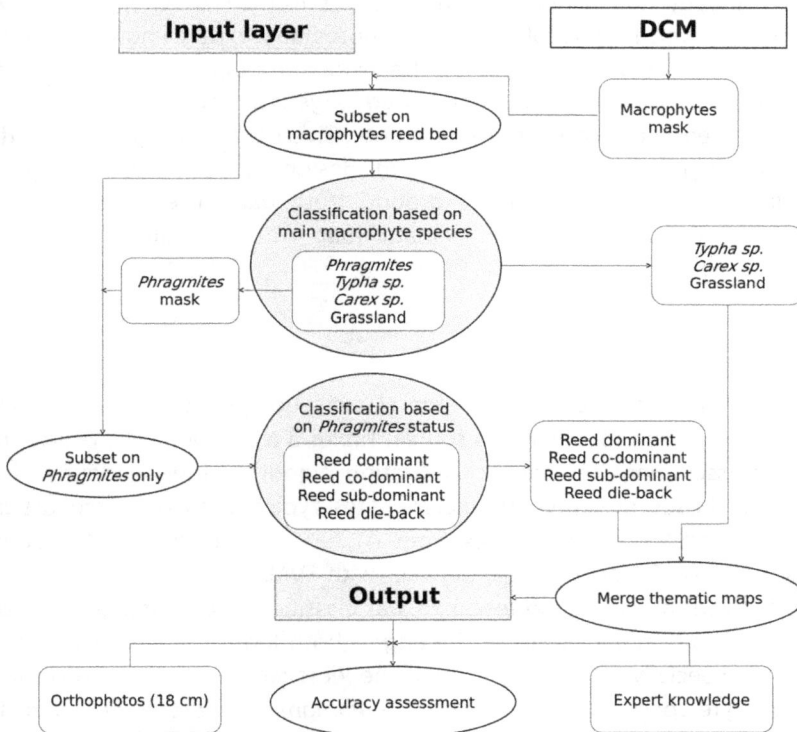

Figure 7. Workflow of the classification scheme developed through iterative classifications and evaluations for mapping emergent vegetation. The two-step approach involves classification of the image based on the main macrophyte species (*Phragmites*, *Typha*, *Carex*, and grassland) and subsequently isolation of the *Phragmites* class and classification of the latter based on the dom-inance of *Phragmites* (dominant, co-dominant, and subdominant) and reed dieback. At the final stage, the two classification results are merged with the *Phragmites*-specific classification overlaid over the macrophyte species map. The result is assessed quantitatively based on expert knowledge and concurrently acquired orthophotos.

subdominant category. Finally, reed dieback is the area where reed degraded naturally and is typically encountered at the waterfront edge of the reed bed, where only *Phragmites* is present.

Subsequently, the two classification products were merged with the subclasses of *Phragmites* substituting the generic class *Phragmites* in the main classification. This scheme was applied twice, first based on the ML algorithm and a second time based on SVM for each input layer. Image processing and classification were realized with the software ENVI 5.0. The cartographic production was carried out in ArcMAP 10.0 (Environmental Systems Research Institute, Inc.).

4.6. *Accuracy assessment*

Classification of remote-sensing data has no credibility unless its accuracy is assessed (Chen et al. 2004). The methodology for accuracy assessment using an error matrix (contingency table) was followed (Congalton 1991). An approach presented by Dalponte, Bruzzone, and Gianelle (2008) in a study employing similar data has been adapted. As already mentioned, *in situ* floristic information synergistically with the concurrently acquired high-resolution orthophotos and expert knowledge were combined to create the validation polygons. The validation polygons were overlaid on the high-resolution orthophotos and were used together with expert knowledge to decide on the vegetation class representation. Given the homogeneity of the polygon, which was taken into account when designing the sample polygons, the ascribed classes were considered the ground truth data. The accuracy assessment was carried out by comparing the thematic map with the validation data using a confusion matrix.

5. Results

Figures 8 and 9 present the results of the classification based on ML and SVM algorithms, respectively, and for each data source. Tables 3 and 4 present the error matrix of the accuracy assessment for each classification algorithm independently. The maps contain solely classes of emergent macrophytes typically encountered around Lake Balaton. The overall accuracy ranges from 41.79% for the lidar data set with ML classification to 88.64% for the Eagle data set with SVM.

Reed in the central part of the reed bed was classified as co-dominant and botanical surveys support this finding, while at the edges (both terrestrial and waterward) of the reed bed and especially in the thin sliver at the west reed bed seems to compete with other macrophyte species. The main class (i.e. *Phragmites*) typically grows in the same environment with other macrophyte species and grasses, and hence the dominant class in all classification results is reed co-dominant which occupies the main reed bed.

Co-dominant reed is very accurately classified by SVM in the cases of Eagle and indices layers with 96% and 97% producer accuracy. Pure reed (i.e. reed dominant) is encountered at a small part of the Lake shore east of the *Typha* island, part which is correctly classified with both ML and SVM in all data sources except lidar; in the case of SVM this part is only slightly containing reed dominant pixels while with ML a large part of the image is classified as reed dominant which is incorrect.

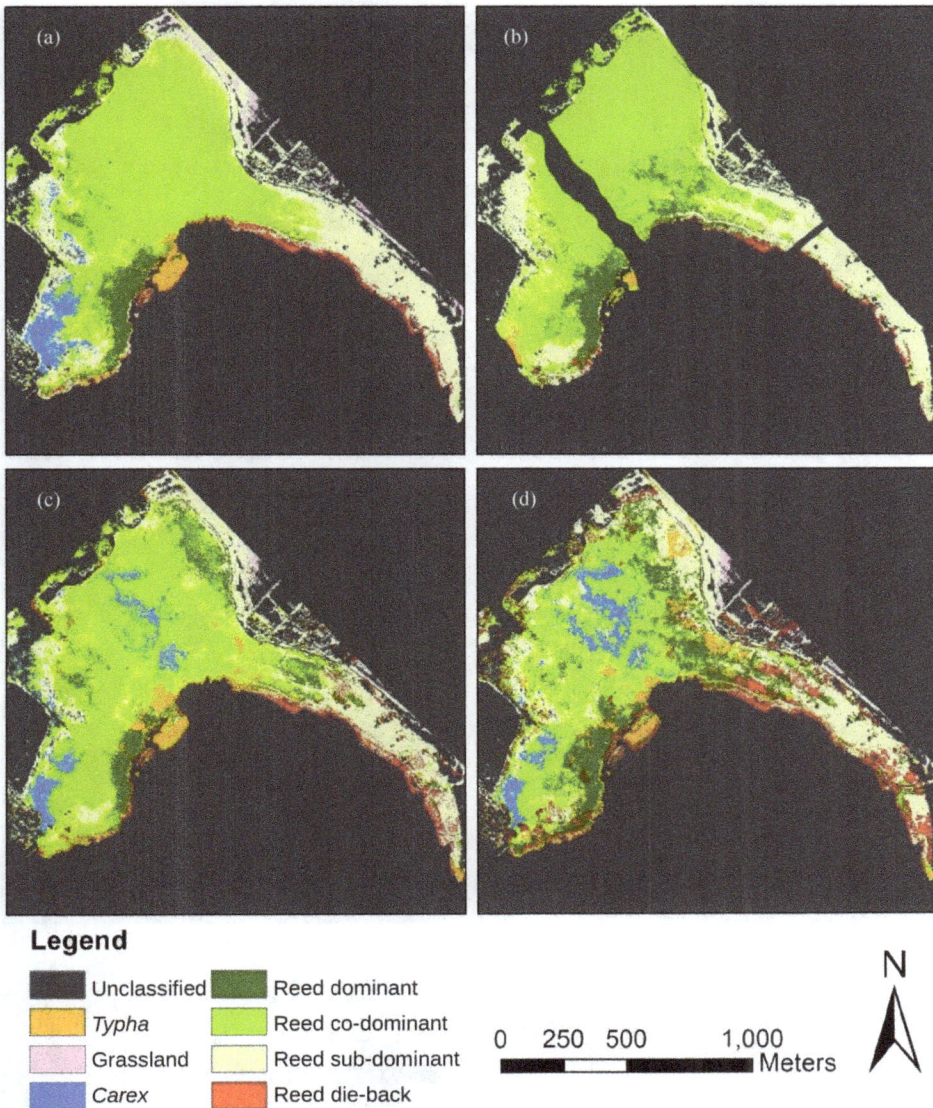

Figure 8. Thematic maps from the ML classification of the Eagle (a), Hawk (b), indices from Eagle (c), and lidar combined with Eagle PCs (d) of the macrophyte main species and *Phragmites* associations at Bozsai Bay, Lake Balaton.

The canopy structure information contained in the lidar does not assist in the classification of this class. SVM shows a higher accuracy for Eagle and indices data sources (84% and 68%, respectively) in comparison to ML; however, when classifying Hawk data ML has an advantage of producer accuracy 60% over 40% for SVM.

It is worth noting that reed dominant has the lowest accuracy across macrophytes as homogeneous and pure patches of *Phragmites* are not frequently encountered in the Bozsai Bay and to an extent they always contain a degree of other macrophyte species within a pixel. Reed subdominant contains reed as a minority within other macrophytes,

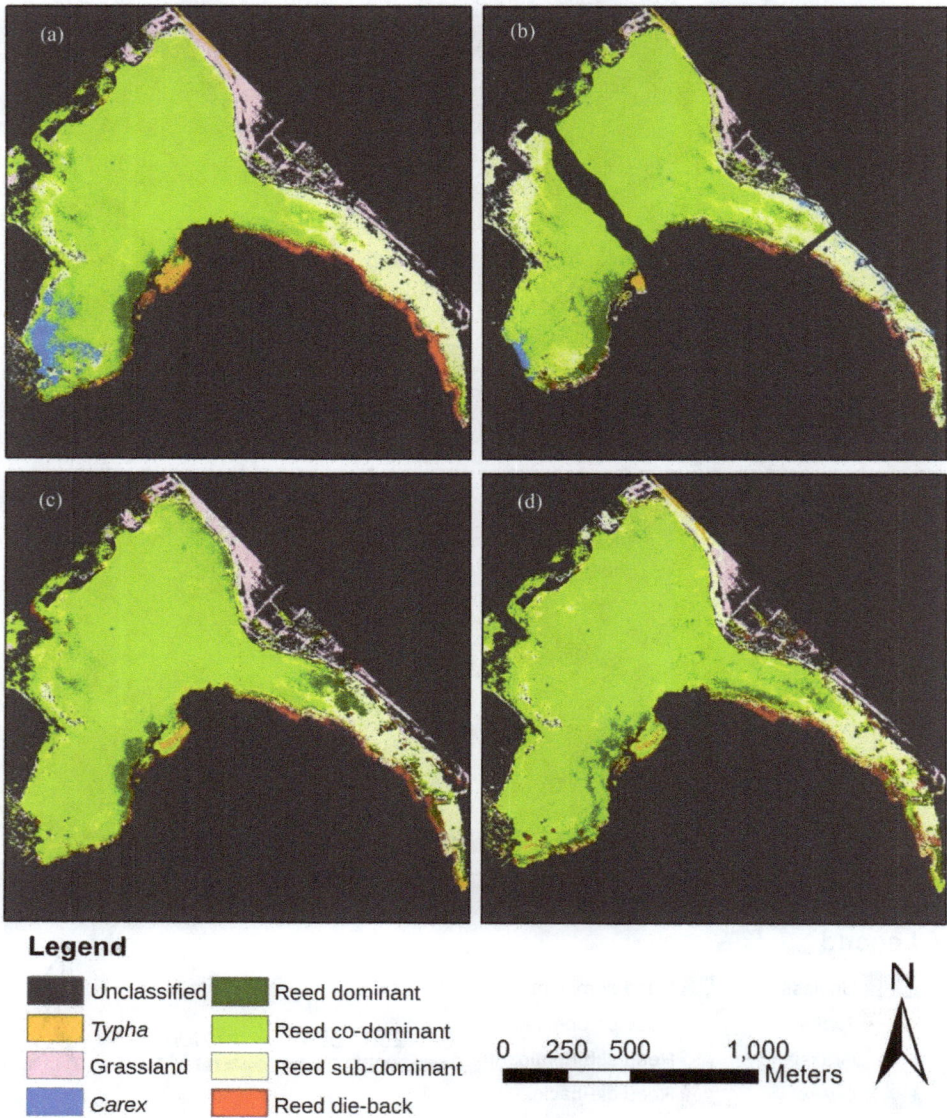

Figure 9. Thematic maps from the SVM classification of the Eagle (a), Hawk (b), indices from Eagle (c), and lidar combined with Eagle PCs (d) of the macrophyte main species and *Phragmites* associations at Bozsai Bay, Lake Balaton.

and is encountered at the edge of the reed bed, where terrestrial vegetation and grass grow in favourable conditions. ML and SVM are classifying the east part of the reed bed as reed subdominant as well as fringe areas of the reed bed. One important difference is found on the north part of the reed bed where there is a sliver of grassland as indicated by SVM classifications of Eagle image and the indices image; however, ML classifies this part mainly as reed subdominant which is incorrect. ML always provides better producer accuracy than SVM, while the opposite is realized for the case of user accuracy; in

Table 3. Confusion matrix illustrating the results of the classification of the four input scenarios with the Maximum Likelihood (ML) algorithm.

	Typha				Grassland				Carex				Reed dominant			
	Eagle	Hawk	Indices	Lidar fusion	Eagle	Hawk	Indices	Lidar fusion	Eagle	Hawk	Indices	Lidar fusion	Eagle	Hawk	Indices	Lidar fusion
Typha	95	63	57	41	–	3	–	1	–	29	–	0	6	0	23	17
Grassland	–	–	–	–	25	–	7	6	–	–	–	–	–	–	–	–
Carex	–	–	–	–	–	–	1	–	95	–	81	73	–	–	–	–
Reed dominant	–	4	7	33	–	–	1	0	–	–	–	0	65	60	55	63
Reed co-dominant	–	29	30	2	3	46	3	1	1	38	13	17	0	14	5	13
Reed sub-dominant	1	4	–	5	69	50	84	92	4	33	6	8	24	13	8	1
Reed dieback	4	–	6	19	2	–	5	–	–	–	–	1	3	12	7	4
Producer accuracy	95	63	57	41	25	–	7	6	95	–	81	73	65	60	55	63
Overall accuracy (%)	83.22	64.44	69.05	41.79												
Kappa coefficient	0.72	0.45	0.52	0.26												

	Reed co-dominant				Reed subdominant				Reed dieback				User accuracy			
	Eagle	Hawk	Indices	Lidar fusion	Eagle	Hawk	Indices	Lidar fusion	Eagle	Hawk	Indices	Lidar fusion	Eagle	Hawk	Indices	Lidar fusion
Typha	0	0	10	11	0	0	2	3	1	–	3	2	81	44	12	9
Grassland	–	–	–	–	–	–	–	–	–	–	–	–	100	–	100	100
Carex	0	–	4	4	3	–	0	1	–	–	2	3	82	–	57	54
Reed dominant	6	22	2	38	0	0	13	17	–	9	1	–	43	15	33	9
Reed co-dominant	86	63	74	31	4	2	8	5	26	5	31	29	98	87	93	88
Reed subdominant	7	15	9	13	89	94	70	65	5	20	–	–	71	63	63	56
Reed dieback	1	–	–	3	2	1	5	7	68	41	58	60	49	43	32	18
Producer accuracy	86	63	74	31	89	94	70	65	68	41	58	60				

All values except the kappa coefficient are in percentages.

Table 4. Confusion matrix illustrating the results of the classification of the four input scenarios with the Support Vector Machines (SVM) algorithm.

	Typha				Grassland				Carex				Reed dominant			
	Eagle	Hawk	Indices	Lidar fusion	Eagle	Hawk	Indices	Lidar fusion	Eagle	Hawk	Indices	Lidar fusion	Eagle	Hawk	Indices	Lidar fusion
Typha	54	59	20	8	–	0	–	–	–	–	–	–	1	0	10	1
Grassland	–	15	–	–	92	69	68	39	–	–	–	–	–	3	–	–
Carex	–	–	–	–	–	–	–	–	84	30	97	95	–	0	–	–
Reed dominant	19	4	9	7	–	26	1	–	–	–	–	–	84	40	68	51
Reed co-dominant	10	20	68	80	6	5	9	4	14	57	3	5	11	42	18	44
Reed subdominant	2	1	–	0	–	–	18	57	2	12	–	–	0	8	0	1
Reed dieback	15	1	3	5	–	–	3	0	–	1	–	–	2	5	2	1
Producer accuracy	54	59	20	8	92	69	68	39	84	30	97	95	84	40	68	51
Overall accuracy (%)	88.64	79.49	77.69	68.72												
Kappa coefficient	0.80	0.65	0.59	0.43												

	Reed co-dominant				Reed subdominant				Reed dieback				User accuracy			
	Eagle	Hawk	Indices	Lidar fusion	Eagle	Hawk	Indices	Lidar fusion	Eagle	Hawk	Indices	Lidar fusion	Eagle	Hawk	Indices	Lidar fusion
Typha	0	0	1	0	1	0	1	2	5	2	3	6	75	79	26	14
Grassland	0	0	0	–	0	1	0	0	1	4	4	0	97	79	97	96
Carex	0	0	–	–	0	4	–	–	–	–	–	–	94	56	–	–
Reed dominant	2	8	0	10	2	1	20	7	3	26	7	6	65	23	34	20
Reed co-dominant	96	85	97	87	14	3	21	35	13	1	1	4	92	89	83	76
Reed subdominant	2	6	1	2	75	86	54	50	6	8	37	28	92	82	84	73
Reed dieback	–	–	0	0	6	2	2	3	72	34	43	49	37	46	45	45
Producer accuracy	96	85	97	87	75	86	54	50	72	34	43	49				

All values except the kappa coefficient are in percentages.

essence this means that SVM results show a realistic map of subdominant reed; however, several pixels from the validation set have not been labelled correctly by SVM and therefore some pixels of this class are misclassified as other classes on this map.

Reed dieback is encountered at the waterward edge of the northeast part of the reed bed mainly as fragmented patches of water reed. Reed dieback is confused mainly with reed subdominant and to a lesser extent with *Typha* as it is mainly neighbouring with the first and the fringe of the *Typha* island is misclassified as dieback in several cases in regard to the latter. This also means that spectrally, and at 1.5 m spatial resolution, dieback is different than the more homogeneous classes of dominant and co-dominant reed. SVM overestimates the extent of reed dieback in the case of Eagle data in comparison to the ML results, while the opposite is the case for the indices composite image. An ML classification of the lidar data presents unrealistic results while the SVM is able to indicate the extent of the dieback reed.

Studies in the literature are controversial about the capability of imaging spectroscopy for mapping vegetation stress. Swatantran et al. (2011), for example, mentioned that hyperspectral data can provide information such as stress on canopy state, while Leckie et al. (2005) attempted unsuccessfully to include unhealthy classes of species in a classification of old growth temperate conifer forest canopies. Zlinszky et al. (2012), using only airborne lidar-based radiometry and structure metrics, reported accuracies of 62% and 76% specifically for reed dieback.

In this study, representative polygons of reed dieback areas were selected; however, reed dieback is a dynamic phenomenon characterizing the physiological condition of a plant and not a vegetation category with concrete boundaries. As such reed dieback areas are indicated by the fragmentation of the reed patch rather than the physiological status of the plant in this classification scheme – this is why lidar proved successful in recognizing the characteristic fragmentation. For a more representative estimation of the physiological status, spectral indices can show the degree to which the area is stable (Stratoulias et al. 2017). However, integrating the dieback class in a macrophyte classification scheme provides an indication of fragmented and sparse reed patches which are under unfavourable environments at the period of image acquisition and hence potentially associated with reed dieback conditions.

Typha, occupying the small island in the centre of the image, is correctly classified in all ML results; however, in the cases of the indices and lidar layers several *Typha* patches are indicated in the reed bed erroneously as indicated by the relatively higher user's accuracy error in these two data sources (Table 3). In the case of the SVM algorithm (Table 4), the Eagle classification provides similar results to the ML results, and the indices and lidar layers indicate the outer buffer zone as reed co-dominant. *Typha* is often confused with other assemblages of *Phragmites*, a fact which is also reported in a similar study from Maheu-Giroux and De Blois (2005) using colour aerial photography and by Zlinszky et al. (2014) using lidar. *Carex* is encountered at the west of the reed bed, classified correctly only in the case of Eagle data for both algorithms with 82% and 94% user accuracy, respectively; however when using the Hawk data, ML does not classify any pixel as *Carex* and instead replaces it with *Typha*. SVM, on the other hand, presents an area restricted in size. In the case of indices and lidar, there is a fragmented distribution in other parts of the image as well as for the ML classification, while SVM does not classify *Carex* pixels. Again the most representative source for pure macro-phytes species is the Eagle instrument.

Grassland is a class differing in foliage height from all the other vegetation types of the study area and hence it would be expected to be distinguished from the lidar composite image; however, this is not the case for ML (producers' accuracy 6%); SVM only partially classifies the sliver on the north part of the image (producers' accuracy 39%). ML in all cases does not perform satisfactorily in the case of grassland as it is most often classified as reed subdominant. This is because the transition from the reed bed to the grassland does not have a crisp boundary but occurs gradually in a buffer zone where the two species coexist in different proportions. SVM applied on the Eagle data set is outperforming considerably for the grass class with 92% producer accuracy and 97% user accuracy.

Regarding the comparison of the two classifiers, overall SVM provides more accurate results than ML, a fact also observed in a similar study mapping submerged macro-phytes by Hunter et al. (2010). However, some errors exist between the different classes of reed and especially the classes encountered at the edge of the reed bed, for instance, reed dieback and sub-dominant reed. Moreover, we observe that ML is not performing well when classifying multisource images while SVM is not confused by this approach. ML seems to underperform when classifying data sets comprising different data types, such as the lidar or the Hawk integrated with the Eagle data. Overall, our study is in agreement with Pal and Mather (2005), who argued that SVM achieves higher level of classification accuracy. This fact can be attributed to two reasons for the case of mapping aquatic vegetation; first, the fine resolution required to map the small extent of a typical aquatic vegetation patch restricts the number, and most importantly the size, of the training data sets to be used in the classification training; second, the hyperspectral data, required to discriminate the complex vegetation status of these types of ecosystems, is introducing high dimensionality in the input data source. Both constrains have proven to be an advantage of the SVM classifier over ML.

While satellite hyperspectral systems are still limited, very fine spatial resolution satellite imagery has been lately available raising similar classification challenges as addressed in this study. Mapping small-scale phenomena, such as lakeshore vegetation, requires a different approach than the application of traditional classifiers (e.g. ML) on traditionally medium-resolution data sets (e.g. Landsat) and therefore the approach to the problem needs to be refined. In this study, we suggest that for fine-scale vegetation mapping the spectral information in the visible and near-infrared provides the most discriminatory power in comparison to longer wavelengths, lidar or the fusion of these data sets. The SVM classifiers are suggested for small-scale phenomena for compensating for the limitation of small training sampling. Potentially, other classification approaches, such as fuzzy classification and machine learning techniques, could provide robust results on different levels of vegetation class mixing encountered in the study area.

6. Conclusions

Classification results derived from ML and SVM classifiers with four different airborne input data sources have been presented. Hyperspectral Eagle and Hawk data have been used independently and synergistically with lidar data to map emergent macrophytes in a nature reserve on the shore of Lake Balaton. A detailed representation of most classes was achieved, which can be attributed to the concurrent very high spectral and spatial resolution of the imagery.

Significant preprocessing is required for the airborne hyperspectral data before the classification. While classification techniques can be automated given the availability of a training dataset, preprocessing is site- and image-specific and the intervention of the user is essential. Field data are an important part of mapping aquatic vegetation; a classification based solely on the airborne imagery and without the contribution of information on the ecological status of the training samples would not provide ecologically meaningful results.

Phragmites is growing in the same environment as other macrophytes and grasses, especially in the terrestrial part of the reed bed. Hence, there is an abundance of reed within any single pixel. In this study, a categorical approach of reed abundance and other macrophyte species as well as grasses was chosen. The reed subclasses show a high degree of separability, with the most prominent class (co-dominant reed) exhibiting 96% producers' accuracy and 92% users' accuracy when an SVM is applied to Eagle data. Reed dieback is the most challenging case of vegetation mapping and unique since it refers to a state of deterioration of the vegetation species rather than species or association of species. Nevertheless, fragmentation at the edge of the reed bed is associated to the consequences of the dieback conditions and hence the extent to which the reed bed is affected.

In agreement with other studies, SVM outperforms ML, mainly for the grassland class, and provides higher overall accuracy with any of the data sources, reaching 89% for optical input data together with near-infrared data. The joint classification of a simple lidar-based canopy model with the first three PCs from the Eagle image did not perform satisfactorily (overall accuracy 69%, $\kappa = 0.43$), which might be attributed to the fact that different macrophytes have a similar canopy structure to be distinguished from a lidar-derived CHM from first and last returns, unlike the high contrast the CHM indicated for more generic classes. However, additional lidar texture indicators such as sigma(z), height variance, or height ranges in a neighbourhood might provide more powerful indicators of vegetation structure (beyond height). Macrophyte associations and species encountered in a typical reed bed on the shore of Lake Balaton are identified using the MNF transformation of high spatial resolution hyperspectral data, especially in the visible and near-infrared spectrum.

The main conclusion of this study is that the classification of solely hyperspectral visible and near-infrared data between 400 and 1000 nm provides the most accurate results in contrast with synergistic classification of hyperspectral data and a CHM or even hyperspectral data extending up to 2500 nm. Moreover, in the case of classification of aquatic resolution at a fine spatial and spectral scale the SVM classifier has overall provided more accurate results in comparison to ML.

Acknowledgments

This work was supported by GIONET, funded by the European Commission, Marie Curie Programme, Initial Training Networks under Grant Agreement PITN-GA-2010-26450. The airborne data were collected and provided by the Airborne Research and Survey Facility vested in the Natural Environment Research Council under the EUFAR Contract Number 2271. AZ was supported by the Hungarian Scientific Research Fund OTKA grant PD 115833. H. Balzter was supported by the Royal Society Wolfson Research Merit Award, 2011/R3 and the NERC National Centre for Earth Observation.

Disclosure statement

No potential conflict of interest was reported by the authors.

Funding

This work was supported by GIONET, funded by the European Commission, Marie Curie Programme, Initial Training Networks under [Grant Agreement PITN-GA-2010-26450]. The airborne data were collected and provided by the Airborne Research and Survey Facility vested in the Natural Environment Research Council under the [EUFAR Contract Number 2271]. AZ was supported by the Hungarian Scientific Research Fund OTKA [grant PD 115833]. H. Balzter was supported by the Royal Society Wolfson Research Merit Award, 2011/R3 and the NERC National Centre for Earth Observation.

ORCID

Dimitris Stratoulias (iD) http://orcid.org/0000-0002-3133-9432

References

Adam, E., O. Mutanga, and D. Rugege. 2010. "Multispectral and Hyperspectral Remote Sensing for Identification and Mapping of Wetland Vegetation: A Review." *Wetland Ecology and Management* 18 (3): 281–296. doi:10.1007/s11273-009-9169-z.

Anderson, K., J. Bennie, and A. Wetherelt. 2010. "Laser Scanning of Fine Scale Pattern along a Hydrological Gradient in a Peatland Ecosystem." *Landscape Ecology* 25: 477–492. doi:10.1007/s10980-009-9408-y.

Artigas, F. J., and J. S. Yang. 2005. "Hyperspectral Remote Sensing of Marsh Species and Plant Vigour Gradient in the New Jersey Meadowlands." *International Journal of Remote Sensing* 26 (23): 5209–5220. doi:10.1080/01431160500218952.

Bahria, S., N. Essoussi, and M. Limam. 2011. "Hyperspectral Data Classification Using Geostatistics and Support Vector Machines." *Remote Sensing Letters* 2: 99–106. doi:10.1080/01431161.2010.497782.

Baker, C., R. Lawrence, C. Montagne, and D. Patten. 2006. "Mapping Wetlands and Riparian Areas Using Landsat ETM+ Imagery and Decision-Tree-Based Models." *Wetlands* 26 (2): 465–474. doi:10.1672/0277-5212(2006)26[465:MWARAU]2.0.CO;2.

Balzter, H., A. Luckman, L. Skinner, C. Rowland, and T. Dawson. 2007. "Observations of Forest Stand Top Height and Mean Height from Interferometric SAR and Lidar over a Conifer Plantation at Thetford Forest, UK." *International Journal of Remote Sensing* 28 (6): 1173–1197. doi:10.1080/01431160600904998.

Bartlett, D. S., and V. Klemas. 1980. "Quantitative Assessment of Tidal Wetlands Using Remote Sensing." *Environmental Management* 4 (4): 337–345. doi:10.1007/BF01869426.

Boyd, D., C. Sanchez-Hernandez, and G. Foody. 2006. "Mapping a Specific Class for Priority Habitats Monitoring from Satellite Sensor Data." *International Journal of Remote Sensing* 27 (13): 2631–2644. doi:10.1080/01431160600554348.

Bradbury, R. B., R. A. Hill, D. C. Mason, S. A. Hinsley, J. D. Wilson, H. Balzter, G. Q. Anderson, M. J. Whittingham, I. J. Davenport, and P. E. Bellamy. 2005. "Modelling Relationships between Birds and Vegetation Structure Using Airborne Lidar Data: A Review with Case Studies from Agricultural and Woodland Environments." *Ibis* 147 (3): 443–452. doi:10.1111/j.1474-919x.2005.00438.x.

Brennan, R., and T. L. Webster. 2006. "Object-Oriented Land Cover Classification of Lidar-Derived Surfaces." *Canadian Journal of Remote Sensing* 32 (2): 162–172. doi:10.5589/m06-015.

Brown, M., H. G. Lewis, and S. R. Gunn. 2000. "Linear Spectral Mixture Models and Support Vector Machines for Remote Sensing." *IEEE Transactions on Geoscience and Remote Sensing* 38 (5): 2346–2360. doi:10.1109/36.868891.

Burai, P., G. Z. Lövei, L. Csaba, I. Nagy, and P. Enyedi. 2010. "Mapping Aquatic Vegetation of the Rakamaz-Tiszanagyfalui Nagy-Morotva Using Hyperspectral Imagery." *AGD Landscape and Environment* 4 (1): 1–10.

Burazerović, D., R. Heylen, B. Geens, S. Sterckx, and P. Scheunders. 2013. "Detecting the Adjacency Effect in Hyperspectral Imagery with Spectral Unmixing Techniques." *IEEE Journal of Selected Topics in Applied Earth Observations and Remote Sensing* 6 (3): 1070–1078. doi:10.1109/JSTARS.2013.2240656.

Butera, K. M. 1983. "Remote Sensing of Wetlands." *IEEE Transactions on Geoscience and Remote Sensing* GE-21 (3): 383–392. doi:10.1109/TGRS.1983.350471.

Camps-Valls, G., and L. Bruzzone. 2005. "Kernel-Based Methods for Hyperspectral Image Classification." *IEEE Transactions on Geoscience and Remote Sensing* 43: 1351–1362. doi:10.1109/TGRS.2005.846154.

Chen, D. M., and D. Stow. 2002. "The Effect of Training Strategies on Supervised Classification at Different Spatial Resolutions." *Photogrammetric Engineering and Remote Sensing* 68 (11): 1155–1161.

Chen, Q., Y. Zhang, A. Ekroos, and M. Hallikainen. 2004. "The Role of Remote Sensing Technology in the EU Water Framework Directive (WFD)." *Environmental Science and Policy* 7: 267–276. doi:10.1016/j.envsci.2004.05.002.

Chi, M., R. Feng, and L. Bruzzone. 2008. "Classification of Hyperspectral Remote-Sensing Data with Primal SVM for Small-Sized Training Dataset Problem." *Advances in Space Research* 41 (11): 1793–1799. doi:10.1016/j.asr.2008.02.012.

Congalton, R. G. 1991. "A Review of Assessing the Accuracy of Classifications of Remotely Sensed Data." *Remote Sensing of Environment* 37 (1): 35–46. doi:10.1016/0034-4257(91)90048-B.

Dalponte, M., L. Bruzzone, and D. Gianelle. 2008. "Fusion of Hyperspectral and Lidar Remote Sensing Data for Classification of Complex Forest Areas." *IEEE Transactions on Geoscience and Remote Sensing* 46 (5): 1416–1427. doi:10.1109/TGRS.2008.916480.

Dalponte, M., L. Bruzzone, L. Vescovo, and D. Gianelle. 2009. "The Role of Spectral Resolution and Classifier Complexity in the Analysis of Hyperspectral Images of Forest Areas." *Remote Sensing of Environment* 113 (11): 2345–2355. doi:10.1016/j.rse.2009.06.013.

Dixon, B., and N. Candade. 2008. "Multispectral Landuse Classification Using Neural Networks and Support Vector Machines: One or the Other, or Both?" *International Journal of Remote Sensing* 29 (4): 1185–1206. doi:10.1080/01431160701294661.

Dvorett, D., C. Davis, and M. Papeş. 2016. "Mapping and Hydrologic Attribution of Temporary Wetlands Using Recurrent Landsat Imagery." *Wetlands* 36 (3): 431–443. doi:10.1007/s13157-016-0752-9.

Eckardt, A., J. Horack, F. Lehmann, D. Krutz, J. Drescher, M. Whorton, and M. Soutullo 2015. Desis (DLR Earth Sensing Imaging Spectrometer for the ISS-MUSES Platform). In *Geoscience and Remote Sensing Symposium (IGARSS), 2015 IEEE International* (pp. 1457–1459). IEEE.

Feingersh, T., and E. B. Dor. 2015. "SHALOM–A Commercial Hyperspectral Space Mission." In *Optical Payloads for Space Missions,* edited by S. Qian, 247. John Wiley & Sons.

Fernandez-Prieto, D., O. Arino, T. Borges, N. Davidson, M. Finlayson, H. Grassl, H. MacKay, C. Prigent, D. Pritchard, and G. Zalidis 2006. The Glob Wetland Symposium: Summary and Way Forward. *Proceedings of the first International Symposium on GlobWetland: Looking at Wetlands from Space,* Frascati, Italy, 19–20 October 2006. ESA SP-634.

Gamon, J. A., J. Penuelas, and C. B. Field. 1992. "A Narrow-Waveband Spectral Index that Tracks Diurnal Changes in Photosynthetic Efficiency." *Remote Sensing of Environment* 41: 35–44. doi:10.1016/0034-4257(92)90059-S.

Gilmore, M. S., D. L. Civco, E. H. Wilson, N. Barrett, S. Prisloe, J. D. Hurd, and C. Chadwick. 2009. "Remote Sensing and in Situ Measurements for Delineation and Assessment of Coastal Marshes and Their Constituent Species." In *Remote Sensing of Coastal Environments*, edited by Y. Wang, 261–280. Boca Raton, Florida: CRC Press.

Gong, P., and P. J. Howarth. 1990. "The Use of Structural Information for Improving Land-Cover Classification Accuracies at the Rural-Urban Fringe." *Photogrammetric Engineering and Remote Sensing* 56 (1): 67–73.

Green, A. A., M. Berman, P. Switzer, and M. D. Craig. 1988. "A Transformation for Ordering Multispectral Data in Terms of Image Quality with Implications for Noise Removal." *IEEE Transactions on Geoscience and Remote Sensing* 26: 65–74. doi:10.1109/36.3001.

Guanter, L., H. Kaufmann, K. Segl, S. Foerster, C. Rogass, S. Chabrillat, ... C. Straif. 2015. "The EnMAP Spaceborne Imaging Spectroscopy Mission for Earth Observation." *Remote Sensing* 7 (7): 8830–8857. doi:10.3390/rs70708830.

Hinsley, S. A., R. A. Hill, P. E. Bellamy, and H. Balzter. 2006. "The Application of Lidar in Woodland Bird Ecology." *Photogrammetric Engineering & Remote Sensing* 72 (12): 1399–1406. doi:10.14358/PERS.72.12.1399.

Hirano, A., M. Madden, and R. Welch. 2003. "Hyperspectral Image Data for Mapping Wetland Vegetation." *Wetlands* 23 (2): 436–448. doi:10.1672/18-20.

Hughes, G. 1968. "On the Mean Accuracy of Statistical Pattern Recognizers." *IEEE Transactions on Information Theory* 14 (1): 55–63. doi:10.1109/TIT.1968.1054102.

Hunter, P. D., D. J. Gilvear, A. N. Tyler, N. J. Willby, and A. Kelly. 2010. "Mapping Macrophytic Vegetation in Shallow Lakes Using the Compact Airborne Spectrographic Imager (CASI)." *Aquatic Conservation-Marine and Freshwater Ecosystems* 20 (7): 717–727. doi:10.1002/aqc.1144.

ITT Visual Information Solutions. 2009. *Atmospheric Correction Module: QUAC and FLAASH User's Guide*. Version 4.7. August. 2009 Edition. ESRI, ENVI.

Jensen, R., P. Mausel, N. Dias, R. Gonser, C. Yang, J. Everitt, and R. Fletcher. 2007. "Spectral Analysis of Coastal Vegetation and Land Cover Using AISA+ Hyperspectral Data." *Geocarto International* 22 (1): 17–28. doi:10.1080/10106040701204354.

Johansen, K., T. Tiede, T. Blaschke, S. Phinn, and L. A. Arroyo 2010. Automatic Geographic Object Based Mapping of Streambed and Riparian Zone Extent from Lidar Data in a Temperate Rural Urban Environment, Australia. *GEOBIA 2010 Geographic Object-Based Image Analysis Conference Proceedings*, XXXVIII-4/C7.

Jones, T. G., N. C. Coops, and T. Sharma. 2010. "Assessing the Utility of Airborne Hyperspectral and LidarData for Species Distribution Mapping in the Coastal Pacific Northwest, Canada." *Remote Sensing of Environment* 114: 2841–2852. doi:10.1016/j.rse.2010.07.002.

Kaplan, G., and U. Avdan. 2017. "Mapping and Monitoring Wetlands Using SENTINEL-2 Satellite Imagery." *ISPRS Annals of Photogrammetry, Remote Sensing and Spatial Information Sciences* 271–277. doi:10.5194/isprs-annals-IV-4-W4-271-2017.

Kashimura, O., K. Hirose, T. Tachikawa, and J. Tanii 2013. Hyperspectral Space-Borne Sensor HISUI and Its Data Application. In *34th Asian Conference on Remote Sensing* (Vol. 2013).

Kaufman, Y. J., D. Tanré, H. R. Gordon, T. Nakajima, J. Lenoble, R. Frouin, H. Grassl, B. M. Herman, M. D. King, and P. M. Teillet. 1997. "Passive Remote Sensing of Tropospheric Aerosol and Atmospheric Correction for the Aerosol Effect." *Journal of Geophysical Research* 102: 16815–16830. doi:10.1029/97JD01496.

Keramitsoglou, I., H. Sarimveis, C. Kiranoudis, C. Kontoes, N. Sifakis, and E. Fitoka. 2006. "The Performance of Pixel Window Algorithms in the Classification of Habitats Using VHSR Imagery." *ISPRS Journal of Photogrammetry and Remote Sensing* 60 (4): 225–238. doi:10.1016/j.isprsjprs.2006.01.002.

Klemas, V. 2011. "Remote Sensing of Wetlands: Case Studies Comparing Practical Techniques." *Journal of Coastal Research* 27 (3): 418–427. doi:10.2112/JCOASTRES-D-10-00174.1.

Koetz, B., F. Morsdorf, T. Curt, S. van der Linden, L. Borgniet, D. Odermatt, S. Alleaume, C. Lampin, M. Jappiot, and B. Allgöwer 2007. Fusion of Imaging Spectrometer and Lidar Data Using Support Vector Machines for Land Cover Classification in the Context of Forest Fire Management. The 10th International Symposium on Physical Measurements and Signatures in Remote Sensing (ISPMSRS), March 12–14, 2007, Davos, Switzerland.

Kovács, M. G., Turcsányi, Z. Tuba, S. E. Wolcsanszky, T. Vasarhelyi, and A. Dely-Draskovits. 1989. "The Decay of Reed in Hungarian Lakes." Symposia Biologica Hungarica 38: 461–471.

Lang, M. W., and G. W. McCarty. 2009. "Lidar Intensity for Improved Detection of Inundation below the Forest Canopy." Wetlands 29 (4): 1166–1178. doi:10.1672/08-197.1.

Lausch, A., M. Pause, I. Merbach, S. Zacharias, D. Doktor, M. Volk, and R. Seppelt. 2013. "A New Multiscale Approach for Monitoring Vegetation Using Remote Sensing-Based Indicators in Laboratory, Field, and Landscape." Environmental Monitoring and Assessment 185: 1215–1235. doi:10.1007/s10661-012-2627-8.

Leckie, D. G., S. Tinis, T. Nelson, C. Burnett, F. A. Gougeon, E. Cloney, and D. Paradine. 2005. "Issues in Species Classification of Trees in Old Growth Conifer Stands." Canadian Journal of Remote Sensing 31 (2): 175–190. doi:10.5589/m05-004.

Lee, C. M., M. L. Cable, S. J. Hook, R. O. Green, S. L. Ustin, D. J. Mandl, and E. M. Middleton. 2015. "An Introduction to the NASA Hyperspectral InfraRed Imager (Hyspiri) Mission and Preparatory Activities." Remote Sensing of Environment 167: 6–19. doi:10.1016/j.rse.2015.06.012.

Lu, D., and Q. Weng. 2007. "A Survey of Image Classification Methods and Techniques for Improving Classification Performance." International Journal of Remote Sensing 28 (5): 823–870. doi:10.1080/01431160600746456.

Maheu-Giroux, M., and S. De Blois. 2005. "Mapping the Invasive Species Phragmites Australis in Linear Wetland Corridors." Aquatic Botany 83 (4): 310–320. doi:10.1016/j.aquabot.2005.07.002.

Malthus, T. J., and D. G. George. 1997. "Airborne Remote Sensing of Macrophytes in Cefni Reservoir, Anglesey, UK." Aquatic Botany 58: 317–332. doi:10.1016/S0304-3770(97)00043-0.

Matthew, M. W., S. M. Adler-Golden, A. Berk, S. C. Richtsmeier, R. Y. Lenive, L. S. Bernstein, P. K. Acharya, et al. 2000. "Status of Atmospheric Correction Using a MODTRAN4-based Algorithm." SPIE Proceedings, Algorithms for Multispectral, Hyperspectral, and Ultraspectral Imagery VI 4049: 199–207.

Maxa, M., and P. Bolstad. 2009. "Mapping Northern Wetlands with High Resolution Satellite Images and Lidar." Wetlands 29 (1): 248–260. doi:10.1672/08-91.1.

Melgani, F., and L. Bruzzone. 2004. "Classification of Hyperspectral Remote Sensing Images with Support Vector Machines." Geoscience and Remote Sensing, IEEE Transactions On 42: 1778–1790. doi:10.1109/TGRS.2004.831865.

Mewes, T., J. Franke, and G. Menz. 2011. "Spectral Requirements on Airborne Hyperspectral Remote Sensing Data for Wheat Disease Detection." Precision Agriculture 12 (6): 795–812. doi:10.1007/s11119-011-9222-9.

Miller, C. 2001. "Fusion of High Resolution Lidar Elevation Data with Hyperspectral Data to Characterize Tree Canopies." Algorithms for Multispectral, Hyperspectral and Ultraspectral Imagery, Vii 4381: 246–252.

Mountrakis, G., J. Im, and C. Ogole. 2011. "Support Vector Machines in Remote Sensing: A Review." ISPRS Journal of Photogrammetry and Remote Sensing 66 (3): 247–259. doi:10.1016/j.isprsjprs.2010.11.001.

NASA LP DAAC. 2001. ASTER L1B. USGS/Earth Resources Observation and Science (EROS) Center. Sioux Falls, South Dakota: NASA.

Neuenschwander, A. L., M. M. Crawford, and M. J. Provancha (1998). Mapping of Coastal Wetlands via Hyperspectral AVIRIS Data. In Geoscience and Remote Sensing Symposium Proceedings, 1998. IGARSS'98. 1998 IEEE International (Vol. 1, pp. 189–191). IEEE.

Niculescu, S., C. Lardeux, I. Grigoras, J. Hanganu, and L. David. 2016. "Synergy between Lidar, RADARSAT-2, and SPOT-5 Images for the Detection and Mapping of Wetland Vegetation in the Danube Delta." IEEE Journal of Selected Topics in Applied Earth Observations and Remote Sensing 9 (8): 3651–3666. doi:10.1109/JSTARS.2016.2545242.

Onojeghuo, A. O., and G. A. Blackburn. 2011. "Optimising the Use of Hyperspectral and LidarData for Mapping Reedbed Habitats." *Remote Sensing of Environment* 115 (8): 2025–2034. doi:10.1016/j.rse.2011.04.004.

Oommen, T., D. Misra, N. K. C. Twarakavi, A. Prakash, B. Sahoo, and S. Bandopadhyay. 2008. "An Objective Analysis of Support Vector Machine Based Classification for Remote Sensing." *Mathematical Geosciences* 40 (4): 409–424. doi:10.1007/s11004-008-9156-6.

Ozesmi, S. L., and M. E. Bauer. 2002. "Satellite Remote Sensing of Wetlands." *Wetlands Ecology and Management* 10 (5): 381–402. doi:10.1023/A:1020908432489.

Pal, M., and M. Mather. 2005. "Support Vector Machines for Classification in Remote Sensing." *International Journal of Remote Sensing* 26 (5): 1007–1011. doi:10.1080/01431160512331314083.

Pal, M., and P. M. Mather. 2004. "Assessment of the Effectiveness of Support Vector Machines for Hyperspectral Data." *Future Generation Computer Systems* 20: 1215–1225. doi:10.1016/j.future.2003.11.011.

Pandey, P. C., N. L. Tate, and H. Balzter. 2014. "Mapping Tree Species in Coastal Portugal Using Statistically Segmented Principal Component Analysis and Other Methods." *IEEE Sensors Journal* 14 (12): 4434–4441. doi:10.1109/JSEN.2014.2335612.

Pedergnana, M., P. R. Marpu, M. Dalla Mura, J. A. Benediktsson, and L. Bruzzone. 2011. "Fusion of Hyperspectral and Lidar Data Using Morphological Attribute Profiles." *Image and Signal Processing for Remote Sensing Xvii* 8180: 81801G.

Pereira, O. J. R., A. J. Melfi, and C. R. Montes. 2017. "Image Fusion of Sentinel-2 and CBERS-4 Satellites for Mapping Soil Cover in The Wetlands of Pantanal." *International Journal of Image and Data Fusion* 8 (2): 148-172. doi:10.1080/19479832.2016.1261946.

Puttonen, E., J. Suomalainen, T. Hakala, E. Raikkonen, H. Kaartinen, S. Kaasalainen, and P. Litkey. 2010. "Tree Species Classification from Fused Active Hyperspectral Reflectance and Lidar Measurements." *Forest Ecology and Management* 260 (10): 1843–1852. doi:10.1016/j.foreco.2010.08.031.

Rebelo, L.-M., C. M. Finlayson, and N. Nagabhatla. 2009. "Remote Sensing and GIS for Wetland Inventory, Mapping and Change Analysis." *Journal of Environmental Management* 90 (7): 2144–2153. doi:10.1016/j.jenvman.2007.06.027.

Reschke, J., and C. Hüttich. 2014. "Continuous Field Mapping of Mediterranean Wetlands Using Sub-Pixel Spectral Signatures and Multi-Temporal Landsat Data." *International Journal of Applied Earth Observation and Geoinformation* 28: 220–229. doi:10.1016/j.jag.2013.12.014.

Richter, R., M. Bachmann, W. Dorigo, and A. Müller. 2006. "Influence of the Adjacency Effect on Ground Reflectance Measurements." *IEEE Geoscience and Remote Sensing Letters* 3 (4): 565–569. doi:10.1109/LGRS.2006.882146.

Robertson, L. D., D. J. King, and C. Davies. 2015. "Assessing Land Cover Change and Anthropogenic Disturbance in Wetlands Using Vegetation Fractions Derived from Landsat 5 TM Imagery (1984–2010)." *Wetlands* 35 (6): 1077–1091. doi:10.1007/s13157-015-0696-5.

Rosso, R. H., S. L. Ustin, and A. Hastings. 2006. "Use of Lidar to Study Changes Associated with Spartina Invasion in San Francisco Bay Marshes." *Remote Sensing of Environment* 100 (3): 295–306. doi:10.1016/j.rse.2005.10.012.

Sanders, L. C., J. R. Schott, and R. Raqueño. 2001. "A VNIR/SWIR Atmospheric Correction Algorithm for Hyperspectral Imagery with Adjacency Effect." *Remote Sensing of Environment* 78 (3): 252–263. doi:10.1016/S0034-4257(01)00219-X.

Santer, R., and C. Schmechtig. 2000. "Adjacency Effects on Water Surfaces: Primary Scattering Approximation and Sensitivity Study." *Applied Optics* 39 (3): 361–375. doi:10.1364/AO.39.000361.

Shafri, H. Z. M., and N. Hamdan. 2009. "Hyperspectral Imagery for Mapping Disease Infection in Oil Palm Plantation Using Vegetation Indices and Red-edge Techniques." *American Journal of Applied Sciences* 6 (6): 1031–1035. doi: 10.3844/ajassp.2009.1031.1035.

Somlyody, L., S. Herodek, and J. Fischer 1983. *Eutrophication of Shallow Lakes: Modeling and Management. The Lake Balaton Case Study*. IIASA Collaborative Paper. IIASA, Laxenburg, Austria, CP-83-703.

Staenz, K., J. Secker, B.-C. Gao, C. Davis, and C. Nadeau. 2002. "Radiative Transfer Codes Applied to Hyperspectral Data for the Retrieval of Surface Reflectance." *ISPRS Journal of Photogrammetry and Remote Sensing* 57 (3): 194–203. doi:10.1016/S0924-2716(02)00121-1.

Stefano, P., P. Angelo, P. Simone, R. Filomena, S. Federico, S. Tiziana, ... M. Stefania 2013. The PRISMA Hyperspectral Mission: Science Activities and Opportunities for Agriculture and Land Monitoring. In *Geoscience and Remote Sensing Symposium (IGARSS), 2013 IEEE International* (pp. 4558–4561). IEEE.

Sterckx, S., E. Knaeps, and K. Ruddick. 2011. "Detection and Correction of Adjacency Effects in Hyperspectral Airborne Data of Coastal and Inland Waters: The Use of the near Infrared Similarity Spectrum." *International Journal of Remote Sensing* 32 (21): 6479–6505. doi:10.1080/01431161.2010.512930.

Stratoulias, D., H. Balzter, A. Zlinszky, and V. R. Tóth. 2015. "Assessment of Ecophysiology of Lake Shore Reed Vegetation Based on Chlorophyll Fluorescence, Field Spectroscopy and Hyperspectral Airborne Imagery." *Remote Sensing of Environment* 157: 72–84. doi:10.1016/j.rse.2014.05.021.

Stratoulias, D., I. Keramitsoglou, P. Burai, L. Csaba, A. Zlinszky, V. R. Tóth, and H. Balzter. 2017. "A Framework for Lakeshore Vegetation Assessment Using Field Spectroscopy and Airborne Hyperspectral Imagery." In *Earth Observation for Land and Emergency Monitoring*, edited by H. Balzter. UK: John Wiley & Sons.

Swatantran, A., R. Dubayah, D. Roberts, M. Hofton, and J. B. Blair. 2011. "Mapping Biomass and Stress in the Sierra Nevada Using Lidar and Hyperspectral Data Fusion." *Remote Sensing of Environment* 115: 2917–2930. doi:10.1016/j.rse.2010.08.027.

Tucker, C. J. 1979. "Red and Photographic Infrared Linear Combinations for Monitoring Vegetation." *Remote Sensing of the Environment* 8: 127–150. doi:10.1016/0034-4257(79)90013-0.

Turner, W., S. Spector, N. Gardiner, M. Fladeland, E. Sterling, and M. Steininger. 2003. "Remote Sensing for Biodiversity Science and Conservation." *Trends in Ecology and Evolution* 18 (6): 306–314. doi:10.1016/S0169-5347(03)00070-3.

Van Der Putten, W. H. 1997. "Die-Back of *Phragmites Australis* in European Wetlands: An Overview of the European Research Programme on Reed Die-Back and Progression (1993–1994)." *Aquatic Botany* 59 (3–4): 263–275. doi:10.1016/S0304-3770(97)00060-0.

Vapnik, V. N. 1995. *The Nature of Statistical Learning Theory*. New York: Springer.

Virág, Á. 1997. *A Balaton Múltja És Jelene*. Eger: Egri Nyomda Rt (In Hungarian).

Warren, M. A., B. H. Taylor, M. G. Grant, and J. D. Shutler. 2014. "Data Processing of Remotely Sensed Airborne Hyperspectral Data Using the Airborne Processing Library (APL): Geocorrection Algorithm Descriptions and Spatial Accuracy Assessment." *Computers & Geosciences* 64: 24-34.

Yang, C., J. A. Goolsby, J. H. Everitt, and Q. Du. 2012. "Applying Six Classifiers to Airborne Hyperspectral Imagery for Detecting Giant Reed." *Geocarto International* 27 (5): 413–424. doi:10.1080/10106049.2011.643321.

Yang, J., and F. J. Artigas. 2009. "Mapping Salt Marsh Vegetation by Integrating Hyperspectral and Lidar Remote Sensing." In *Remote Sensing of Coastal Environments*, edited by Y. Wang. US: CRC Press.

Zlinszky, A., A. Schroiff, A. Kania, B. Deák, W. Mücke, Á. Vári, B. Székely, and N. Pfeifer. 2014. "Categorizing Grassland Vegetation with Full-waveform Airborne Laser Scanning: a Feasibility Study for Detecting Natura 2000 Habitat Types." *Remote Sensing* 6 (9): 8056-8087. doi:10.3390/rs6098056.

Zlinszky, A., V. R. Tóth, P. Pomogyi, and G. Timár. 2011. "Initial Report of the AIMWETLAND Project: Simultaneous Airborne Hyperspectral, Lidar and Photogrammetric Survey of the Full Shoreline of Lake Balaton, Hungary." *Geographia Technica* 1: 101–117.

Zlinszky, A., W. Mücke, H. Lehner, C. Briese, and N. Pfeifer. 2012. "Categorizing Wetland Vegetation by Airborne Laser Scanning on Lake Balaton and Kis-Balaton, Hungary." *Remote Sensing* 4 (6): 1617–1650. doi:10.3390/rs4061617.

Predicting macroalgal pigments (chlorophyll *a*, chlorophyll *b*, chlorophyll *a* + *b*, carotenoids) in various environmental conditions using high-resolution hyperspectral spectroradiometers

Ele Vahtmäe, Jonne Kotta, Helen Orav-Kotta, Ilmar Kotta, Merli Pärnoja and Tiit Kutser

ABSTRACT

Photosynthetic pigments may indicate the health and productivity of vegetation and thereby are among the most important targets of the remote-sensing science. We studied the relationship between macroalgae pigment concentration measured *in situ* and spectral reflectance, to develop predictive remote-sensing methods for macroalgal pigments. The measurements of spectral reflectance of macroalgae were made using both a field portable spectrometer Ramses built by TriOS GmbH (Germany) and a laboratory hyperspectral imaging device HySpex built by Norsk Elektro Optikk (Norway). Our results showed that differences in total chlorophyll (Chl-a + b) concentrations resulted in the consistent change of spectral reflectance for studied brown (*Fucus vesiculosus*) and green (*Cladophora glomerata*, *Ulva intestinalis*) macroalgae species. Charophytes (*Chara aspera*, *Chara horrida*) were also studied, and the relationship was much weaker for this taxon. If spectral indices predicted relatively well the concentration of Chl-a + b (R^2 = 0.64–0.73) and the carotenoid to total chlorophyll ratio (Car:Chl-a + b, R^2 = 0.80) across the five studied macroalgae species, then the concentration of chlorophyll a (Chl-a), chlorophyll b (Chl-b), and carotenoids (Car) were more difficult to model (R^2 = 0.004–0.51). The HySpex imaging system yielded systematically better results in predicting pigment concentrations compared to the Ramses spectroradiometer. By using traditional assessment of pigment concentration along with the Hyspex imaging device, we were able to build models with a capability to predict the spatial patterns of pigment concentration for Baltic Sea macroalgae.

1. Introduction

The health of vegetation is related to the existence of photosynthesizing pigments. Determination of the presence and density of plant pigments is important in the assessment of terrestrial vegetation as well as benthic macroalgae health and productivity. There is a strong interest today in developing and validating techniques to detect and quantify individual plants pigments that can advance our understanding of biophysical functioning in plants (Ustin et al. 2006).

Within a species, a variety of factors, including growth stage and various environmental conditions, can change the total pigment content and the pigment ratio (Anderson, Chow, and Godchild 1988). Pigmentation can be directly related to stress physiology, as concentrations of carotenoids (Car) increase and chlorophylls (Chl) generally decrease under environmental stress and during senescence (Peñuelas and Filella 1998). An understanding of plant responses to fluctuations in environment is critical to predict how plants and ecosystems respond to climate change (Chapin, Rincon, and Huante 1993).

There is also a close relationship between a plant's pigment content, photosynthetic potential, and primary production. The amount of solar radiation absorbed by a leaf is largely a function of the foliar concentrations of photosynthetic pigments, and therefore, low concentrations of Chl can directly limit photosynthetic potential and hence primary production (Richardson, Duigan, and Berlyn 2002). Detecting photosynthetic rates and gross primary production is essential for evaluating the global carbon cycle for research on climate change (Garbulsky et al. 2014).

Traditional methods for measuring pigment concentration involve extraction with a solvent and subsequent spectrophotometric or high-performance liquid chromatography analysis using standard laboratory procedures (Blackburn 2007). Although those methods are reliable, they are destructive, time consuming, labour demanding, and expensive (Richardson, Duigan, and Berlyn 2002).

Faster non-destructive measurements of pigment concentration can be obtained using optical spectroradiometers (Blackburn 1998; Carter and Knapp 2001; Richardson, Duigan, and Berlyn 2002). Hyperspectral spectroradiometers can potentially detect quite small changes in a plant's biochemicals and hence rather subtle changes that are often characteristic of the early effects of stress (Jones and Vaughan 2010). To date, there exists a large number of studies that relate reflectance spectra with the pigment concentration in terrestrial plants (Chappelle, Kim, and McMurtrey 1992; Vogelmann, Rock, and Moss 1993; Peñuelas, Baret, and Filella 1995; Gitelson and Merzlyak 1996, 1997; Blackburn 1998; Carter and Knapp 2001; Richardson, Duigan, and Berlyn 2002; Sims and Gamon 2002; LeMaire, Francois, and Dufrene 2004). Much less is known about coral reef habitats (Torres-Perez et al. 2015; Joyce and Phinn 2003). The latter studies have also included macroalgae at some extent, but we could not find any comprehensive research that relates macroalgal reflectance spectra to their pigments.

Most studies that quantify plant pigments by optical spectroradiometers have used non-imaging spectral measurements, which provide information about a limited surface area of the plant. For example, sensors with fibre optics provide information about a single point (1 mm^2) on the plant's surface. Sensors without fibre optics provide information about larger areas, but the measured spectral information is averaged over the studied area. The studied area itself depends on the sensors field of view (FOV) as well as the distance between the sensor and the object. These non-imaging measurements are not sufficient to obtain the spatial distribution of pigments in the whole plant or larger part of the plant (Jones and Vaughan 2010). It is important to know the distribution of pigments as structural and functional changes in plants do not take place simultaneously throughout the whole tissue but are patchy in their distribution (Bergsträsser et al. 2015). Hyperspectral imaging sensors are able to acquire both spectral and spatial information simultaneously and thereby allow determination of the spatial distribution of pigments.

Macroalgae species have habitat requirements that are closely linked to surrounding water quality conditions. Light, temperature, nutrients, water movement, and salinity primarily control the photosynthetic production of marine macroalgae (Kirst 1990). Located at the margins of typical marine environments, the Baltic Sea is a vulnerable ecosystem and strong gradients in salinity and temperature challenge macroalgae species (Kotta et al. 2014). Moreover, the Northern Hemisphere high-latitude regions are expected to experience more severe global warming compared to low-latitude regions (IPCC 2013; Koch et al. 2013). The most important potential future climate stressors in the Baltic Sea are increases in temperature and nutrients load, and decrease in salinity (Andersson et al. 2015; Meier et al. 2012; Bring, Rogberg, and Destouni 2015). These profound alterations in Baltic Sea environmental conditions will likely lead to significant changes in macroalgal physiology. Linking ambient stresses and pigments in aquatic macroalgae under current and future environmental conditions allows us to understand how the Baltic Sea macroalgae can cope with these stresses.

The main aim of the current study was to assess the relationship between spectral reflectance of Baltic Sea macroalgae and their pigment composition (Chl-a, Chl-b, Chl-$a + b$, Car) with the purpose of predicting macroalgal pigment concentration from remote-sensing data. The main objectives were (1) to study the relationship between macroalgal pigment concentration and spectral reflectance, (2) to compare the capability of non-imaging Ramses and imaging HySpex spectroradiometers in predicting macroalgal pigment concentration (species level and across species) using spectral indices, (3) to predict the spatial distribution of pigment concentration using imaging spectroradiometer, and (4) to study how environmental variables relevant to future climate change scenarios of the Baltic Sea region influence macroalgal pigment concentration.

2. Material and methods

The flow chart of the methodology used in the current study is presented in Figure 1. Macroalgae samples were collected from the Baltic Sea coastal water and incubated for 2 weeks under the most plausible future environmental conditions. The measurements of reflectance spectra of incubated macroalgae were made using first Ramses non-imaging and then HySpex imaging spectrometer. On completion of spectral reflectance measurements, pigment concentrations of macroalgae were measured in laboratory. Different published spectral indices were calculated from the Ramses and HySpex reflectance data. Regression analyses were conducted to assess the relationship between measured pigment concentrations and spectral indices separately for the Ramses and HySpex data. The best-performing regression equations retrieved by correlating spectral indices and pigment concentrations were applied to the HySpex image to examine the spatial distribution of pigments.

2.1. Macroalgae and sampling site

The macroalgae species used in this study were those commonly found in the shallow water habitat of the northeastern Baltic Sea. Live and healthy specimens were collected

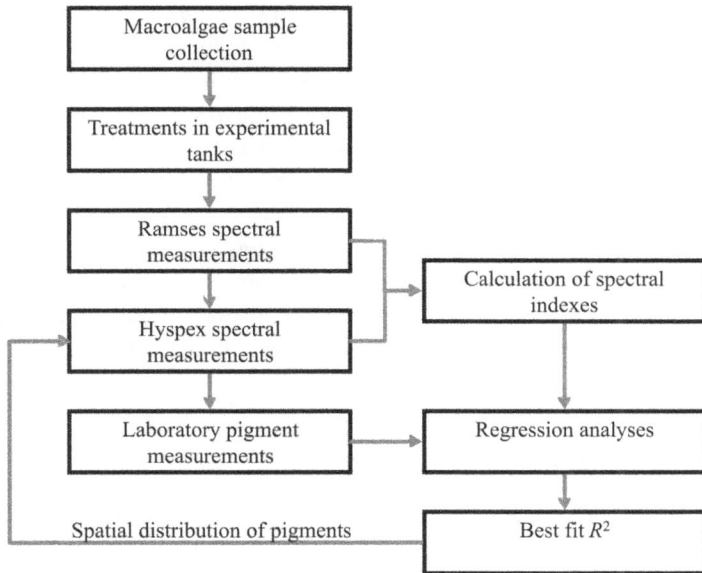

Figure 1. The flow chart of methodology used in the current study.

from Kõiguste Bay and the Small Strait at two individual sampling dates in summer (26.07.2016) and autumn (21.09.2016). The collected specimens were immediately transported to the experimental rooms in the Kõiguste Field Station. The brown macroalga *Fucus vesiculosus*, the green macroalga *Cladophora glomerata* and the charophytes *Chara aspera* and *Chara horrida* were collected in summer season. The green macroalga *C. glomerata* was replaced by *Ulva intestinalis* in autumn as the healthy *C. glomerata* is not normally found in this season. Species that grow attached to boulders (e.g. *F. vesiculosus*, *C. glomerata*, *U. intestinalis*) were collected together with boulders.

The Baltic Sea populations of *F. vesiculosus* have broad temperature optima for growth between 10°C and 24°C with highest rates estimated at 15–20°C. Optimal temperature for photosynthesis is higher (estimated at 24°C) and much narrower compared to that for growth (Graiff et al. 2015). *F. vesiculosus* can be found from the southern Baltic Sea at salinity close to ocean water to the Gulf of Bothnia at salinity down to 4 psu (Kautsky 1992; Raven and Samuelsson 1988) and in the Gulf of Finland down to 3–6 psu (Back and Ruuskanen 2000).

C. glomerata and *U. intestinalis* are morphologically similar filamentous green algae species, which occupy a similar niche in the upper coastal zone of the brackish Baltic Sea (Choo et al. 2005). *C. glomerata* is a freshwater species, but it is also widely spread in the brackish water conditions of the Baltic Sea. *C. glomerata* is sensitive to saline waters with its net photosynthesis being significantly reduced at salinities of 11 psu and above (Thomas, Collins, and Russell 1988). *C. glomerata* is photosynthetically most active at high temperatures (20–30°C) and does not tolerate very low temperatures. *U. intestinalis* is a marine species (Alström-Rapaport, Leskinen, and Pamilo 2010). However, as *U. intestinalis* has a broad salinity tolerance, the species is found all over the brackish Baltic Sea except in the innermost bays in the Gulf of Finland and the Bothnian Bay where salinity values drop down to 2 psu (Leskinen, Alström-Rapaport, and Pamilo 2004).

Chara species are common in freshwater environments. Among the Baltic Sea charophytes, *C. aspera* covers the widest salinity range (Blindow 2000) with upper salinity limits near 14 psu (Torn 2008). There is no evidence that the temperature regime has any strong influence on charophyte populations although several investigations show that temperature affects the photosynthetic performance of charophytes (Torn and Martin 2004).

The nutrient load is known to strongly correlate with the biomass of macroalgae. In general, elevated nutrient loads into the northeastern Baltic Sea increases the species richness of macrophytes and the biomass of annual macroalgae (e.g. *C. glomerata, U. intestinalis*). However, perennial macroalgae (*F. vesiculosus*) are depressed (Kotta and Möller 2014).

2.2. *Experimental design*

The most plausible future environment of the Baltic Sea was simulated under simple experimental conditions. In summer, water tanks of 60 l volume were prepared and deployed outdoors under natural daylight and temperature (18°C). In these tanks, four different environmental conditions were generated: (S) natural sea water collected from Kõiguste Bay and representing reference conditions of the current environment (6–7 psu), (S + F) reduced salinity condition, which was produced by mixing equal amounts of the bay water and fresh water, (S + N) natural sea water enriched with nutrients twice as high as natural background, (S + F + N) reduced salinity enriched with nutrients. Elevated nutrient conditions of these latter treatments were obtained by adding fertilizer sticks to the water.

In autumn, similar experiment was performed but all these four treatments were run at two temperature conditions: natural ambient (13°C) (S, S + F, S + N, S + F + N) and increased temperature conditions (18°C) (S°, S + F°, S + N°, S + F + N°). The elevated temperature conditions were obtained by placing water tanks in a climate controlled room with the light regime similar to field conditions. In summer, higher-than-normal temperature conditions were not mimicked as the climate change phenomena of the northeastern Baltic Sea region are represented by elevated temperatures in cold seasons and heat waves in summer are very unlikely to happen.

All water tanks were equipped with an air stone connected to a self-priming pump in order to provide permanent water circulation. For each macroalgae species, three specimens were deployed at each experimental condition for 2 weeks.

2.3. *Reflectance measurements*

The measurements of spectral reflectance of macroalgae were made using both a field portable spectrometer (Ramses, built by TriOS GmbH, Germany) and a laboratory hyperspectral imaging device (HySpex, built by Norsk Elektro Optikk, NEO, Norway). The productive parts of the incubated macroalgae were used in the measurements.

The Ramses spectrometer is a non-imaging sensor allowing measurements of a single spectral signal at a time. The Ramses optical system was composed of two simultaneously operated sensors: irradiance and radiance sensors. Downwelling light (E_d, W m^{-2} nm^{-1}) was measured by the irradiance sensor and upwelling light (L_u, W m^{-2} nm^{-1} sr^{-1}) by the radiance sensor. The reflectance spectrum of each sample was calculated as a ratio of upwelling radiance to downwelling irradiance (L_u/E_d). The FOV of the Ramses radiance sensor is 7°. The upwelling light was measured above the sample from 10 cm distance resulting in a measurement area of 1.1 cm^2. Ramses sensors cover the visible and near-infrared range of the spectrum from 350 to 900 nm with about a 3 nm spectral interval. During the measurements, macroalgae samples were placed on a black plastic bag to minimize any signal from the adjacent environment. The measurements were performed under natural light conditions. Five to ten individual spectra were measured for each sample and a mean spectrum was calculated. After the Ramses measurements, macroalgae samples were put back to the seawater for the following HySpex measurements.

In the laboratory, the HySpex hyperspectral imaging sensor was mounted on a scanning device. The device allows scanning of the whole plant for reflectance measurements determining the spatial distribution of spectral information. The scanning device is equipped with a certified Tungsten Halogen light source fitted with a heat filter that absorbs light strongly at wavelengths above 700 nm and absorbs light completely at 850 and longer wavelengths. This light source was selected to cause minimal thermal stress on the samples during measurements. However, the test measurements showed that this lamp does not provide sufficient amount of energy at 750 nm and longer wavelength that are often used to estimate the concentration of Chl. Therefore, we decided to use an alternative light source, a regular Tungsten lamp providing sufficient amount of radiation in the near-infrared (NIR) and infrared (IR) part of the spectrum.

The HySpex sensor covers the spectral range from 410 to 988 nm at a sampling interval of 2.7 nm and provides spectral information in 216 bands. Macroalgal samples were blotted dry, placed on a black background, and scanned from 30 cm distance at constant speed (30 mm min^{-1}) to acquire raw spectral data for each pixel (0.1 mm spatial resolution with 30 cm lense). Samples were kept beneath the light source for the minimum time required to scan the sample in order to avoid excessive drying out of the samples. It is unlikely that any significant pigment degradation took place over that time to cause any problems in interpretation of the results. Spectralon panel with 10% reflectivity was used for reference measurements after each sample scan. Collected raw HySpex data were converted into units of spectral radiance (W m^{-2} nm^{-1} sr^{-1}) using software developed by the NEO.

On completion of spectral reflectance measurements, macroalgal samples were immediately wrapped using aluminium foil and frozen for laboratory pigment analysis.

Subsequent image analysis was conducted using the ENVI software (Research System Inc., Boulder CO, USA). The radiance values for each imaged pixel were standardized to reflectance spectra by dividing sample spectra by the average reference spectra of Spectralon panel. Data were filtered by low pass filter in order to reduce the instrument noise.

2.4. Spectral indices

In spectroscopy, spectral indices have been developed to reduce complex reflectance spectra to a single value (Richardson, Duigan, and Berlyn 2002). Different published spectral indices were tested for pigment retrieval. These included various single band, ratio-based as well as normalized ratio-based indices with different wavelengths, red

Table 1. Published spectral indices tested in this study.

Spectral index	Equation	Pigment	References
SR	R_{750}/R_{705}	Chl-a + b	Jordan (1969
NDVI	$(R_{750} - R_{705})/(R_{750} + R_{705})$	Chl-a + b	Gitelson and Merzlyak (1994, 1996)
mSR	$(R_{750} - R_{445})/(R_{705} - R_{445})$	Chl-a + b	Sims and Gamon (2002)
mNDVI	$(R_{750} - R_{705})/(R_{750} + R_{705} - 2x\ R_{445})$	Chl-a + b	Sims and Gamon (2002)
RE3/RE2	$(\text{mean}[R_{734};R_{747}]/\text{mean}[R_{715};R_{726}])$	Chl-a + b	Vogelmann, Rock, and Moss (1993)
PSSRa	R_{800}/R_{680}	Chl-a	Blackburn (1998)
PSSRb	R_{800}/R_{635}	Chl-b	Blackburn (1998)
PSSRc	R_{800}/R_{500}	Car	Blackburn (1998)
PRI	$(R_{531} - R_{570})/(R_{531} + R_{570})$	Car:Chl-a + b	Gamon, Peñuelas, and Field (1992)

edge features, and derivative values. Out of these indices, five were used in further data analysis as they showed higher correlations with the pigment concentrations of all the studied macroalgae species (Table 1).

These were (1) simple ratio (SR), (2) normalized difference vegetation index (NDVI), (3) RE$_3$/RE$_2$, (4) pigment-specific simple ratios (PSSR), and (5) photochemical reflectance index (PRI). We used the nearest wavelength if exact wavelength proposed in the literature was absent from sensor's available bands.

SR index was used for Chl-a + b retrieval (Jordan 1969). Most ratio-based indices employ ratios of narrow bands within areas of the spectrum that are sensitive to pigments and those areas not sensitive to pigments (Blackburn 2007). An SR index for Chl retrieval typically divides reflectance at a reference wavelength (typically between 750 and 900 nm) by an index wavelength (typically between 660 and 720 nm) (Sims and Gamon 2002). In this article, we used 750 as a reference wavelength and 705 as an index wavelength.

NDVI was used for Chl-a + b retrieval (Gitelson and Merzlyak 1994, 1996). Indices based on NDVI use the same wavelengths as the SR, but subtract index wavelength from reference wavelength and the value is then normalized through division by the sum of the reflectance at the same two wavelengths.

Sims and Gamon (2002) suggested making further modifications of these indices to compensate for high leaf surface (specular) reflectance, which tends to increase reflectance across the whole visible spectrum. A wavelength of 445 nm was used for the calculation of two modified indices (mSR and mNDVI) as a measure of surface reflection.

Different red edge parameters have been shown to correlate well with variations in Chl-a + b content. Loss of Chl increases reflectance across the visible and near-infrared spectrum and shifts the red edge (the long-wavelength edge of the Chl absorption) towards shorter wavelengths (Ustin et al. 2004). In the current work, we used RE3/RE2 index, which was calculated as an average reflectance from 734 to 747 nm divided by average reflectance from 715 to 726 nm (Vogelmann, Rock, and Moss 1993).

There have been few investigations of the possibility of estimating the concentrations of photosynthetic pigments other than Chl-a + b from measurements of spectral reflectance. Blackburn (1998) proposed PSSR for Chl-a, Chl-b, and Car. The basis of PSSR approach was to develop a simple spectral index for each pigment of interest, using a similar structure to that of the SR index. Each index uses a near-infrared band (800 nm) as a reference, which can be considered to minimize the effects of radiation interactions at the leaf surface and internal structures in the mesophyll, as suggested by

Peñuelas et al. (1995). A wavelength of 680 nm was used for Chl-*a* retrieval, which is the absorption peak of Chl-*a*. For retrieval of Chl-*b*, 635 and 650 nm have been suggested by different authors (Chappelle, Kim, and McMurtrey 1992; Blackburn 1998; Richardson, Duigan, and Berlyn 2002). We used 635 nm, which has been reported by Blackburn (1998) to be the optimum wavelength best correlating with Chl-*b* content. Wavelengths of 470 and 500 nm have both been suggested for the Car retrieval, but in our case, 500 nm showed slightly better results.

An estimation of leaf Car content from reflectance is much more difficult than the estimation of Chl (Sims and Gamon 2002). Significant overlap in the absorption features of Chl and Car and the low concentration of Car with respect to Chl presents difficulties in defining suitable spectral indices for Car (Blackburn 2007). The estimation of the ratio of Car to Chl has shown to be more successful than estimation of the absolute Car content (Merzlyak et al. 1999; Peñuelas, Baret, and Filella 1995). The PRI was originally developed by Gamon, Peñuelas, and Field (1992) as an indicator of photosynthetic efficiency.

The PRI measures the relative reflectance on either side of the green reflectance peak (550 nm); so, it also compares the reflectance in the blue (Chl and Car absorption) region of the spectrum with the reflectance in the red (Chl absorption only) region (Peñuelas, Garbulsky, and Filella 2011). Consequently, it can serve as an estimate of the Car to Chl-*a* + *b* ratio.

2.5. *Laboratory pigment measurements*

In the laboratory, the region of collected tissue for pigment analysis was marked in the corresponding HySpex scan for future reference (Figure 2). For example, vegetative apical tips (the actively growing regions) were collected in the case of *F. vesiculosus* and the whorls of branchlets were collected in the case of *C. horrida*. Due to the homogeneity of the study objects, it was not possible to identify the exact locations of tissue collection in the case of *C. glomerata*, *U. intestinalis*, and *C. aspera*. In these cases, we collected the tissue from the central regions of the samples. The areas of tissue collection were called the regions of interest (ROIs).

Samples were first washed and excess water was removed by paper towel. Collected tissue (0.1–0.2 g fresh weight) was weighed, cut into small pieces, crushed using mortar and pestle to form homogenous slurry, and mixed with 10 ml methanol (100% pure solvent). Sand was used for plant tissue maceration in case of tougher tissues. The mixture was placed in a refrigerator for 24 h. The samples were then centrifuged and analysed using the PerkinElmer Lambda 35 UV/VIS spectrophotometer. All procedures were carried out under low-light conditions in the laboratory in to minimize photo-oxidation of pigments.

From the spectrophotometer data, absorbance values at 665.2, 652.4, and 470 nm were used to determine the concentrations of the extracted pigments (Chl-*a*, Chl-*b*, Car) using the equations of Lichtenthaler (1987),

$$C_{Chl\_a} = 16.72A_{665.2} - 9.16A_{652.4}$$
$$C_{Chl\_b} = 34.09A_{652.4} - 15.28A_{665.2}$$
$$C_{Car} = (1000A_{470} - 1.63C_{Chl\_a} - 104.96C_{Chl\_b})/221$$

(1)

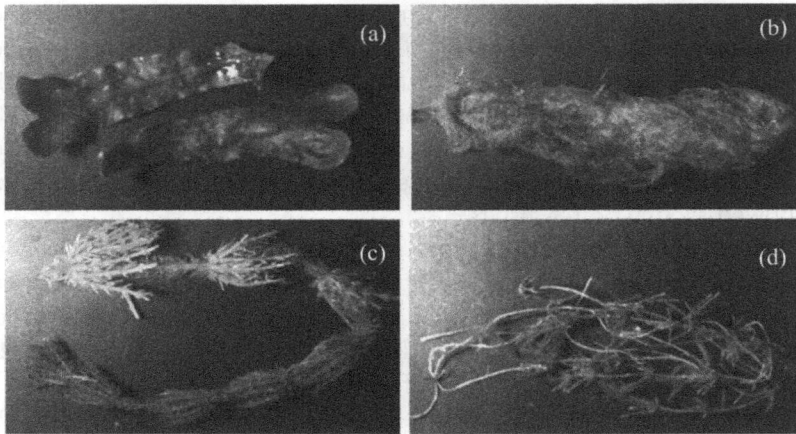

Figure 2. Examples of HySpex scans for (a) *F. vesiculosus*, (b) *C. glomerata*, (c) *C. horrida*, (d) *C. aspera*. The locations of tissue collection were identified as the regions of interests (ROIs) in the HySpex images concurrently with the tissue collection.

These equations gave pigment concentrations µg ml^{-1} of extract, which were then converted to mg g^{-1} wet weight.

2.6. *Data analyses*

The spectra of each pixel in ROIs in HySpex images were averaged and the mean spectrum was considered as the spectrum of the sample. Mean spectra of all samples were then extracted from ROIs in the hyperspectral images.

Reflectance spectra of three samples from the same experimental treatment were averaged resulting in 12 mean reflectance spectra for each macroalgae species. As green macroalga was represented by two different, but morphologically similar species (*C. glomerata* and *U. intestinalis*), they were pooled together and named as green filamentous macroalgae in our data analysis. The mean treatment spectrum was calculated separately both for the Ramses and HySpex measurements. Similarly, a mean pigment concentration as an average of pigment extraction from three corresponding samples from the same experimental treatment was used in the following data analysis. Two green filamentous macroalgae samples and one *F. vesiculosus* sample were considered as outliers and removed from further data analysis.

Spectral indices were calculated from the Ramses and HySpex reflectance data. Regression analyses were conducted to assess the relatedness between measured pigment concentrations and spectral indices separately for the Ramses and HySpex data. The coefficients of determination (R^2) were used to evaluate the strength of relationship between plant pigments and spectral indices.

The best-performing regression equations retrieved by correlating spectral indices and pigment concentrations were applied to the HySpex image to examine the spatial distribution of pigments.

3. Results

3.1. *Linking spectral reflectance to macroalgal pigment concentration*

Figure 3 shows differences in the measured reflectance spectra of studied macroalgae species among different Chl-a + b concentrations. Here, the mean HySpex spectra of each experimental treatment are shown and corresponding mean Chl-a + b concentrations presented. As brown macroalga do not contain Chl-b, the total Chl in *F. vesiculosus* was considered equal to Chl-a. In order to allow better comparison between different spectra, reflectance spectra were normalized based on the mean and standard deviation of the spectra (Figure 3(a)). The first derivative spectra were calculated as the slope of the original reflectance spectra (Figure 3(b)), where shifts in the position of the red edge with changing Chl-a + b are presented.

The measured spectra of brown macroalga *F. vesiculosus* showed the best correspondence with changing Chl-a concentrations. The higher the Chl-a concentration, the broader and flatter was the absorption peak near 680 nm. Moreover, reduced Chl-a concentrations shifted the red to near-infrared transition spectrum towards blue end of the spectrum. Shifts in the position of the red edge with changing Chl-a concentration are especially obvious in the first derivative spectra.

Spectra of green filamentous macroalgae showed also a good relationship to Chl-a + b concentration. The relatedness was much weaker in case of charophytes *C. aspera* and *C. horrida*. The normalized reflectance spectra of charophytes show a rather uniform pattern near 680 nm. The shift of the position of red edge is taking place, but the change does not correspond to Chl-a + b change.

3.2. *The capability of Ramses and HySpex spectroradiometers to predict macroalgal pigment concentration using spectral indices*

Calculated spectral indices were initially regressed against the concentration of macroalgal pigments for each species individually. After individual analyses, we also pooled all studied species into the same regression analysis (combined-1). However, as the pigment concentration of *C. horrida* only very poorly correlated with the used spectral indices, we also ran a separate set of analysis, from which *C. horrida* was excluded (combined-2).

Table 2 compares the strength of relationship between the used spectral indices and pigment concentrations measured by the HySpex imaging spectrometer in terms of the coefficient of determination (R^2). In most cases, the linear regression model performed the best. However, a logarithmic relationship showed the highest correlation between Chl-a + b and NDVI as well as mNDVI indices in charophytes (*C. aspera*, *C. horrida*) and in the combined-1 species model.

The spectral indices calculated from the reflectance spectra of the brown macroalga *F. vesiculosus* and green filamentous macroalgae showed very strong relationship with Chl-a + b. For the charophyte *C. aspera*, the relationship was weaker, displaying higher scatter around the regression line. The spectral indices of the charophyte *C. horrida* did not have linkage with Chl-a + b. When all species were combined together, the relationships between Chl-a + b and used spectral indices remained moderately strong

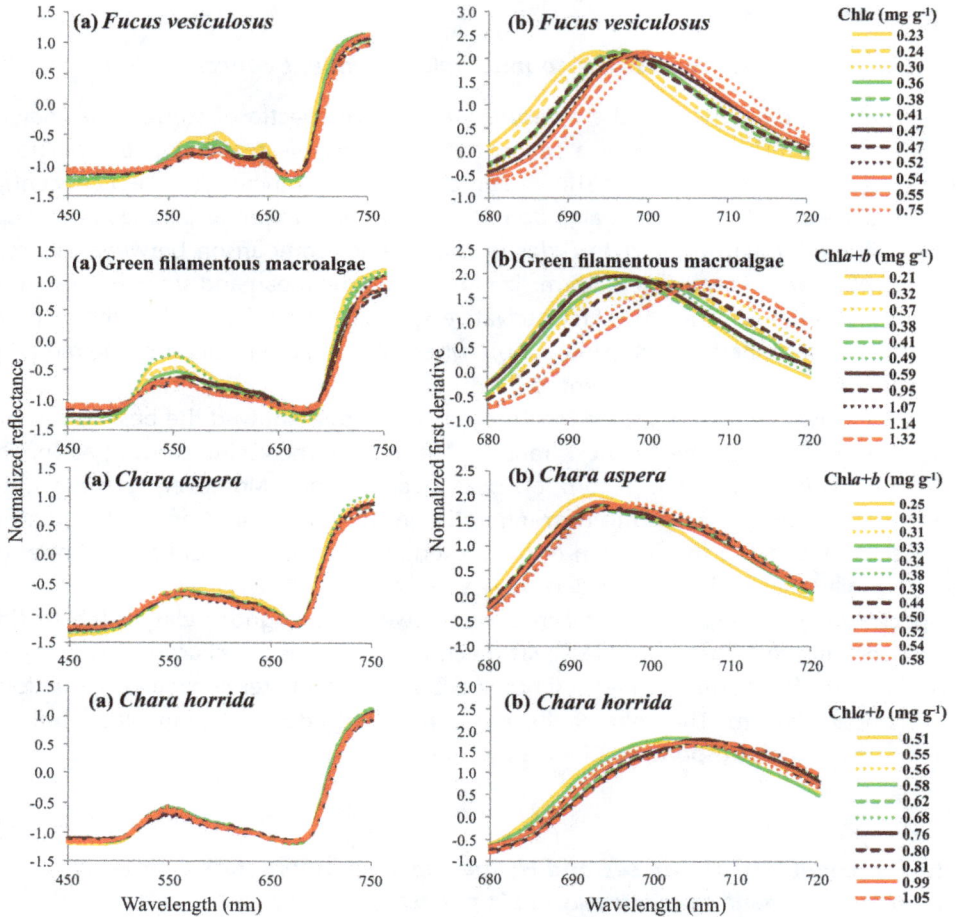

Figure 3. Reflectance spectra of studied macroalgae indicating changes in (a) the normalized reflectance, (b) the normalized first derivative (slope of the original reflectance spectrum) across a range of total chlorophyll (Chl-a + b) concentrations (mg g^{-1}). As brown macroalgae do not contain Chl-b, the total chlorophyll in *F. vesiculosus* was considered equal to Chl-a. Each curve represents the average spectrum of three specimens incubated under the same experimental conditions and measured with the HySpex scanning device. Total chlorophyll concentrations indicate the mean measured chlorophyll concentration of the same three specimens.

(R^2 = 0.64–0.73). The similar relationships involving combined-2 species were somewhat higher (R^2 = 0.77–0.85).

Figure 4 plots the indices that had the strongest relationships with Chl-a + b concentrations in each case (different macroalgae species and combined species). SR index showed highest correlation with Chl-a + b in the model of brown macroalga *F. vesiculosus* (R^2 = 0.88). The best-performing index for Chl-a + b retrieval in the green filamentous macroalgae was RE3/RE2 (R^2 = 0.81). RE3/RE2 index performed also the best in case of the charophyte *C. aspera* (R^2 = 0.73) and the combined-2 species model (R^2 = 0.85). The mNDVI index performed best correlating with Chl-a + b in the combined-1 species model (R^2 = 0.73).

Table 2. The coefficients of determination (R^2) between spectral indices and pigment concentrations in the HySpex spectral measurements.

Spectral index	Related to	Fucus vesiculosus R^2	Green filamentous macroalgae R^2	Chara aspera R^2	Chara horrida R^2	Combined-1 all species R^2	Combined-2 excl. Chara horrida R^2
SR	Chl-a + b	0.88	0.76	0.60	0.03	0.64	0.80
NDVI	Chl-a + b	0.87	0.74	0.64	0.07	0.69	0.77
mSR	Chl-a + b	0.86	0.80	0.67	0.10	0.69	0.82
mNDVI	Chl-a + b	0.86	0.78	0.70	0.16	0.73	0.79
RE3/RE2	Chl-a + b	0.86	0.81	0.73	0.11	0.68	0.85
PSSRa	Chl-a	0.77	0.34	0.58	0.24	0.26	0.40
PSSRb	Chl-b	–	0.73	0.63	0.03	0.51 [a]	0.78 [a]
PSSRc	Car	0.60	0.23	0.13	0.06	<0.01	0.19
PRI	Car:Chl	<0.01	0.51	0.69	0.22	0.80	0.76

Chl-a + b, Chl-a, Chl-b, Car are the concentrations of total chlorophyll, chlorophyll a, chlorophyll b, and carotenoids. [a]Except F. vesiculosus.

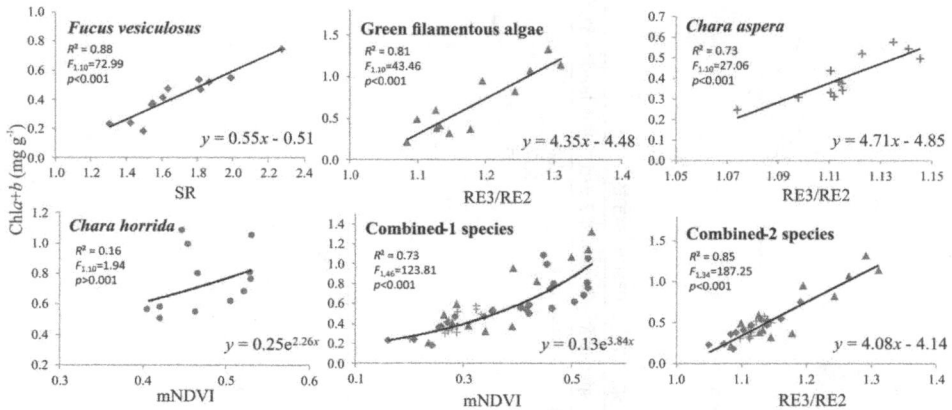

Figure 4. Best performing models between total chlorophyll (Chl-a + b, mg g^{-1} wet wt) concentration and spectral indices separately for different macroalgal species and in combined species models. Spectral indices were calculated from the HySpex measurements. The Combined-1 species model incorporates all studied macroalgae species and the combined-2 species model incorporates all macroalgae species except *C. horrida*.

Except for *F. vesiculosus*, PSSRa and PSSRc only poorly correlated with Chl-a and Car. PSSRb had better match with Chl-b. The linkage between PRI and Car:Chl-a + b ratio varied largely among different macroalgae species. However, PRI correlated very well with Car:Chl-a + b ratio in the combined-1 species model (R^2 = 0.80) and the combined-2 species model (R^2 = 0.76).

Figure 5 displays regressions between spectral indices and Chl-a, Chl-b, and Car:Chl-a + b ratio for combined-1 species. PSSRa produced rather low regression with Chl-a (R^2 = 0.26), while PSSRb performed significantly better in models including Chl-b (R^2 = 0.51). PRI showed high performance in predicting the Car:Chl-a + b ratio in the model combining all macroalgae species (R^2 = 0.80).

Table 3 compares the strength of relationship between the used spectral indices and pigment concentrations measured by the Ramses imaging spectrometer in terms of the coefficient of determination (R^2). Similar to the HySpex measurements, linear relationship between pigments and indices mostly provided the best results. Logarithmic relationships showed the highest linkage between Chl-a + b and NDVI as well as mNDVI indices in green filamentous macroalgae, charophytes (*C. aspera, C. horrida*),

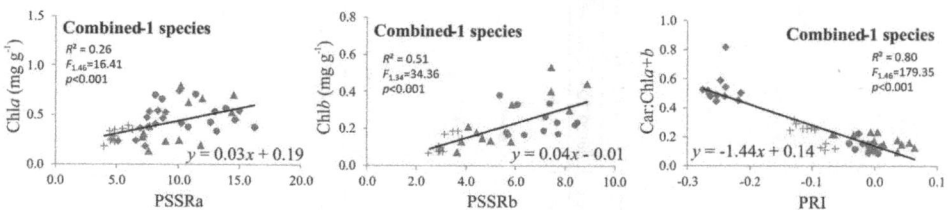

Figure 5. Regressions between Chl-a, Chl-b, Car:Chl-a + b ratio and spectral indices in models incorporating all studied macroalgae species (combined-1 species). Spectral indices were calculated from the HySpex measurements.

Table 3. The coefficients of determination (R^2) between spectral indices and pigment concentrations in the Ramses spectral measurements.

Spectral index	Related to	Fucus vesiculosus R^2	Green filamentous macroalgae R^2	Chara aspera R^2	Chara horrida R^2	Combined-1 all species R^2	Combined-2 excl. Chara horrida R^2
SR	Chl-a + b	0.68	0.56	0.09	<0.01	0.44	0.58
NDVI	Chl-a + b	0.64	0.52	0.16	<0.01	0.49	0.52
mSR	Chl-a + b	0.57	0.80	0.48	0.03	0.56	0.76
mNDVI	Chl-a + b	0.78	0.82	0.58	0.03	0.67	0.76
RE3/RE2	Chl-a + b	0.55	0.57	0.12	0.01	0.41	0.55
PSSRa	Chl-a	0.21	0.02	<0.01	<0.01	0.13	0.15
PSSRb	Chl-b	–	0.37	<0.01	<0.01	0.32[a]	0.51[a]
PSSRc	Car	<0.01	<0.01	0.40	0.02	0.05	0.10
PRI	Car:Chl	0.08	0.45	0.73	0.36	0.81	0.79

Chl-a + b, Chl-a, Chl-b, Car are concentrations of total chlorophyll, chlorophyll a, chlorophyll b, and carotenoids. [a]Except F. vesiculosus.

the combined-1 species, and the combined-2 species models. With some exceptions, the strength of the relationship was much lower in the Ramses measurements compared to the HySpex measurements.

The Chl-a + b concentrations of the brown macroalga F. vesiculosus and green filamentous macroalgae were moderately linked with spectral indices. A weaker match was found in case of the carophyte C. aspera. Once again, the Chl-a + b concentration of the charophyte C. horrida was practically not linked to any used spectral indices. When all species were combined, the relationships (R^2) between Chl-a + b and spectral indices remained between 0.41 and 0.67 depending on the index. The similar relationships involving combined-2 species were somewhat higher (R^2 = 0.52–0.76).

In the Ramses measurements, Chl-a + b linked better to the modified spectral indices (mSR, mNDVI) than SR and NDVI indices. The mNDVI index had the best predictive capability in Chl-a + b estimation in separate and combined species models.

The PSSR indices did not correlate with Chl-a, Chl-b, and Car. PRI showed relatively high correlations with Car:Chl-a + b ratio outperforming even the results of the HySpex measurements.

3.3. *The spatial distribution of pigment concentration*

The best-performing regression equations relating spectral indices and pigment concentrations were used to predict the spatial patterns of pigments of the studied macroalgae in the HySpex images at pixel level. As we were not able to use a leaf clip to flatten the macroalgae samples while acquiring the images, the assessment of pigments may have some inconsistencies due to shadows.

Figure 6 illustrates the spatial variability in Chl-a for the two F. vesiculosus specimens, both incubated under different experimental conditions. As the SR spectral index showed the highest predictive potential in retrieving the Chl-a concentration of F. vesiculosus, this index was used for the assessment of Chl-a in F. vesiculosus. The ROI areas are marked with red circles on the images.

The laboratory estimate of Chl-a concentration was 0.37 mg g^{-1} for the first specimen and 0.91 mg g^{-1} for the second specimen. The predictions of pigment concentration show that the first F. vesiculosus specimen displays lower Chl-a concentrations between

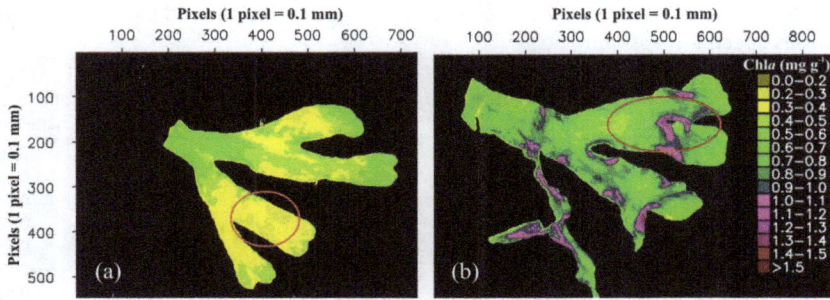

Figure 6. The spatial variability in Chl-*a* concentration of the two specimen of the brown macroalga *F. vesiculosus* (mg g^{-1} wet weight) estimated using the SR index. (a) The macroalga has been incubated under reduced salinity and nutrient enriched water in ambient temperature condition in autumn (S + F + N), (b) the macroalga has been incubated under elevated temperature and nutrient enriched conditions in autumn (S + N°). The ROI areas are marked with red circles.

0.2 and 0.8 mg g^{-1} with the mean ROI value at 0.32 mg g^{-1} and for the second specimen, the values are higher between 0.3 and 1.5 mg g^{-1} with the mean ROI value at 0.77 mg g^{-1}. Such comparison demonstrates that the estimates of Chl-*a* spatial patterns from spectral indices are aligned with the results of laboratory Chl-*a* measurements.

The spatial variability in Chl-*a* + *b* for the two *C. glomerata* specimens is illustrated in Figure 7. RE3/RE2 index was used for Chl-*a* + *b* determination. Due to the homogeneity of the study object, it was not possible to identify the exact locations of tissue collection

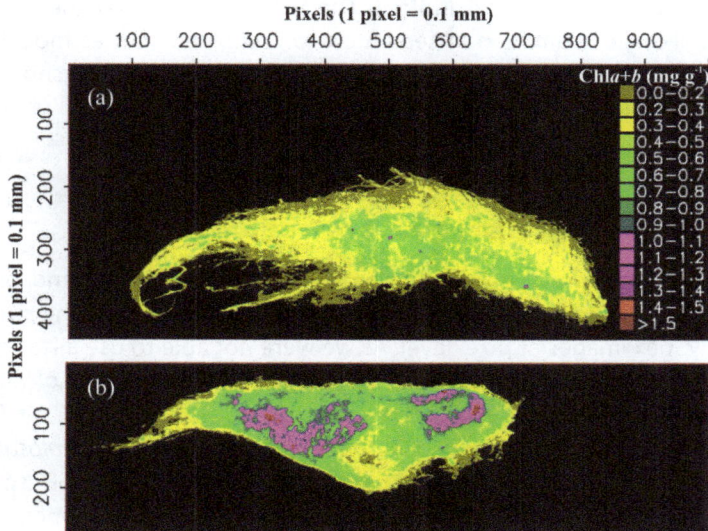

Figure 7. The spatial variability in Chl-*a* + *b* concentration of the two specimen of the green macroalga *C. glomerata* (mg g^{-1} wet weight) estimated using the RE3/RE2 index. (a) The macroalga has been incubated in sea water under ambient temperature condition in summer (S), (b) the macroalga has been incubated under reduced salinity and nutrient enriched water under ambient temperature condition in summer (S + F + N).

in the case of *C. glomerata*, and the tissue was collected from central regions of the samples for the laboratory measurements. The laboratory estimate of Chl-*a* + *b* concentration was 0.32 mg g^{-1} for the first specimen and 1.36 mg g^{-1} for the second specimen. The predictions of pigment concentration show that the first *C. glomerata* specimen displays lower Chl-*a* + *b* concentrations between 0.1 and 1.0 mg g^{-1} with the mean ROI value at around 0.3–0.4 mg g^{-1}. The values are higher for the second specimen reaching more than 1.5 mg g^{-1} with the mean ROI value at around 0.7–0.8 mg g^{-1}.

The HySpex imagery reveals that the spatial patterns of pigments are highly heterogeneous within and between specimens, the later often reflecting exposure to different environmental conditions.

3.4. *Macroalgal pigment concentrations in various environmental conditions*

Figure 8 shows the mean concentrations of Chl-*a*, Chl-*b*, and Car in studied macroalgae under different experimental conditions. As brown macroalga is known to lack Chl-*b*, only Chl-*a* and Car concentrations are shown for *F. vesiculosus*. Our results showed that green filamentous macroalga was the most affected by environmental variations as it had the highest variability in pigment concentration (0.14–0.80 mg g^{-1} for Chl-*a*, 0.07–0.53 mg g^{-1} for Chl-*b*, and 0.05–0.20 mg g^{-1} for Car). The brown macroalga *F. vesiculosus* showed also relatively high variability (0.23–0.75 mg g^{-1} for Chl-*a* and 0.12–0.33 mg g^{-1}

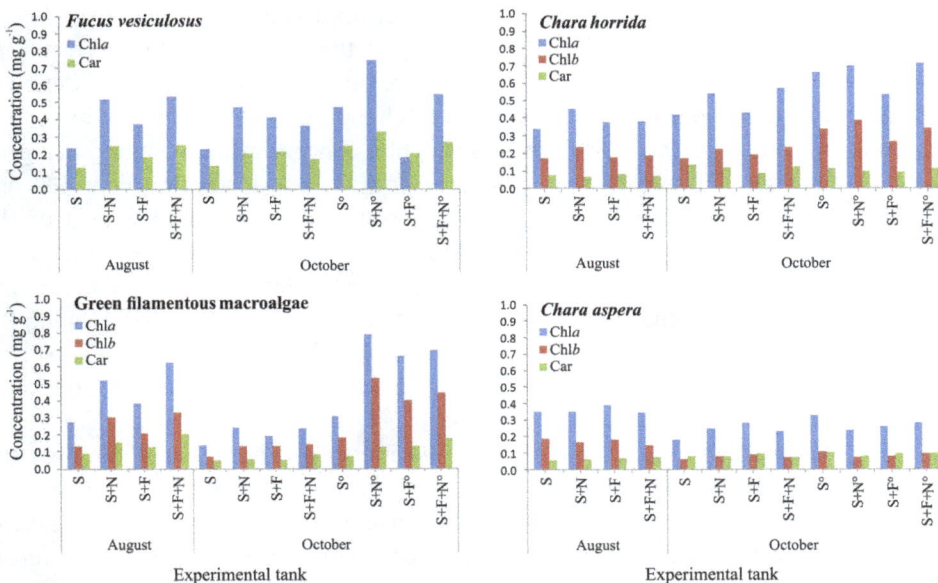

Figure 8. The mean pigment concentrations (mg g^{-1} wet weight) in the studied macroalgae incubated at different experimental conditions: (S) sea water, (S + N) sea water enriched with nutrients, (S + F) reduced salinity, (S + F + N) reduced salinity enriched with nutrients. The degree symbol (°) denotes the elevated temperature condition in autumn. For further experimental details, see Section 2. Chl-*a*, Chl-*b*, Car are the concentration of chlorophyll *a*, chlorophyll *b*, and carotenoids, respectively. As brown macroalgae are known to lack Chl-*b*, only Chl-*a* and Car concentrations are shown for *F. vesiculosus*.

for Car). The pigment concentrations of charophyte *C. horrida* varied two times less, ranging from 0.34 to 0.71 mg g^{-1} for Chl-*a*, from 0.17 to 0.39 mg g^{-1} for Chl-*b*, and from 0.06 to 0.13 mg g^{-1} for Car. The lowest pigment variability was detected in *C. aspera*, where pigment concentrations only ranged from 0.18 to 0.39 mg g^{-1} for Chl-*a*, from 0.06 to 0.19 mg g^{-1} for Chl-*b*, and from 0.06 to 0.11 mg g^{-1} for Car.

In general, all studied macroalgae species, except *C. aspera*, had the highest concentration of Chl-*a* and Chl-*b* in autumn at elevated temperature conditions. *C. aspera* had systematically higher concentration of Chl in summer irrespective of salinity and nutrient conditions. The outdoor incubations in autumn generally resulted the lowest concentration of Chl for all the macroalgae, except for *C. horrida*, which showed systematically lower concentrations in August. In general, nutrient enrichment increased the concentration of Chl-*a* and Chl-*b* except for *C. aspera* that showed the opposite trend. Reduced salinity increased the concentration of Chl-*a* and Chl-*b* under natural environmental conditions in summer and autumn but lowered the values under elevated temperature conditions in autumn. Car concentration did not follow the same trends as Chl.

4. Discussion

In our study, we prepared simple experiments to expose different macroalgae species under a wide range of environmental conditions simulating current and future climate scenarios. After the deployment, we quantified the spectral reflectance and pigment composition of these macroalgae with an aim to seek if and how the spectral information can be used to predict pigment composition of macroalgae.

The first objective of the study was to investigate the relationship between macroalgae pigment composition and spectral reflectance. Our results showed that differences in Chl-*a* concentration resulted in the consistent change of spectral signatures for the brown macroalga *F. vesiculosus*, with decreasing Chl-*a* concentrations resulting in higher reflectance between 550 and 650 nm, narrower and sharper absorption near 680 nm, and the shift of red to near-infrared transition spectrum towards blue end of the spectrum. The same trend was true for the green filamentous macroalgae and the charophyte *C. aspera*, although not expressed as consistently as by *F. vesiculosus*. The reflectance spectra of the charophyte *C. horrida* did not show much variability along the studied gradient of Chl-*a* + *b* concentrations, plausibly due to low relative variability in Chl-*a* + *b* concentration and/or internal structural characteristics related to leaf thickness or water content (Sims and Gamon 2002).

The second objective involved comparing the capability of non-imaging Ramses and imaging HySpex spectroradiometers to predict macroalgae pigment concentration using spectral indices. In the case of the Ramses reflectance measurements, a single measurement averaged over the studied area of more than 1 cm^2 was considered as representative for the whole sample. This method assumed the homogeneity of distribution of pigments across the entire macroalga.

The HySpex imaging system allowed acquiring spectral information with 0.1 mm spatial resolution. However, it was not possible to obtain information about the concentration of pigments of each HySpex pixel by traditional laboratory methods. Therefore, in the case of HySpex measurements, we also used an average reflectance spectrum to represent the sample. However, we used an average value of a small defined part of the sample (ROI region) to better match reflectance measurements and laboratory pigment analyses.

Although ROI regions were chosen from visually homogeneous macroalgae regions, the variability within the ROI region may still have influenced the error in our predictive equations. Moreover, we used an average reflectance and average pigment concentrations of three macroalgae specimens to build our calibration models. Here, significant within-species variability may have lead to confusion in pigment assessment. Nevertheless, our models predicted relatively well the concentration of pigments in macroalgal thalli with the Hyspex imaging system yielding systematically better results compared to the Ramses non-imaging instrument.

Different spectral indices have been more or less successful when relating reflectance spectra to pigment concentration. Many authors have reported that relationships between Chl and spectral indices largely vary among different species, and such differences may severely impair our ability to use these indices across a wide range of vegetation types (e.g. Richardson, Duigan, and Berlyn 2002). Our results showed that many indices reflected well the concentration of Chl-a + b in $F.$ $vesiculosus$, $C.$ $glomerata$, $U.$ $intestinalis$, $C.$ $aspera$, but not in $C.$ $horrida$. Nevertheless, most of the Chl-a + b indices, especially when measured with the HySpex spectrometer, allowed relatively well accurate prediction of the concentration of pigments across the five studied macroalgae species.

The RE3/RE2 index had the best predictive capability in Chl-a + b estimation in the HySpex measurements. This index employs wavelengths in the far-red spectral region from 715 to 726 nm and from 734 to 747 nm. The far-red optical response to stress is explained by the tendency of stressed leaves to lose Chl and the absorption properties of Chl (Carter and Knapp 2001). Although not used very extensively in the literature, the current index has shown relatively high performance in two comprehensive studies aiming at determining best spectral indices over different vegetation species (LeMarie, Francois, and Dufrene 2004; Croft, Chen, and Zhang 2014). In addition, the RE3/RE2 index is simple (Vogelmann, Rock, and Moss 1993) and opposes to many other red edge-related indices that require derivative calculations.

In the Ramses measurements, the RE3/RE2 index showed low performance. Here, modified spectral indices (mSR, mNDVI) correlated better with Chl-a + b than SR and NDVI. From this, we may conclude that the reflectance of plant surface was the most important factor for low index performance as those modified indices are designated to eliminate the effect of such reflectance. The above-mentioned pattern was not detected in the HySpex measurements.

The PSSRa and PSSRb indices were able to predict the concentration of Chl-a and Chl-b to some extent, but no good relationship was found between PSSRc and the concentration of Car. These results are in accordance with the previous findings that the estimation of Car remains more difficult than that of Chl concentration (Sims and Gamon 2002; Blackburn 2007). Cars persist longer than Chl in senescing leaves. Peñuelas et al. (1995) showed that the ratio of Cars:Chl generally rose in senescing and unhealthy plants and decreased in healthy growing plants. The same trend was followed in the current study. We found a generic relationship between PRI and Car:Chl-a + b ratio across all studied macroalgae species.

The next objective of the current study was to predict the spatial distribution of pigment concentration using an imaging spectroradiometer. The development of new imaging systems provides an opportunity to investigate the physiological impacts of environmental factors on the biota and to study if such effects are expressed in a spatially variable fashion (Nicotra et al. 2003). Combining traditional assessment of pigment concentration and hyperspectral imaging, we were able to build empirical models with a capability to predict the spatial patterns of pigment concentration in macroalgae species. The results demonstrated that the spatial distribution of Chl-*a* + *b* was not uniform in the macroalgal surface area.

The final objective was to study how environmental variables relevant to future climate change scenarios of the Baltic Sea region influence macroalgae pigment composition. Our results showed that elevated nutrient load resulted in higher Chl-*a* concentration of the brown macroalga *F. vesiculosus*. This is also in agreement with the findings of Nygard and Dring (2008) who showed that a high nutrient content increased the photosynthesis and growth rate of the Baltic Sea *F. vesiculosus*. *F. vesiculosus* also showed slightly higher pigment concentrations at elevated temperature conditions; nevertheless, all incubations were performed within the optimal temperature range of the species. *F. vesiculosus* is a marine species, but it has well adapted to very low salinities of the Baltic Sea (3–4 psu). Our study showed multifaceted relationship between salinity, temperature, and pigment concentration. Specifically, reduced salinity mostly increased the pigment concentration of *F. vesiculosus* at normal ambient temperature whereas elevated temperature reversed the effect.

Our experiments also showed that temperature was the most important factor that regulated the pigment concentration in the studied green filamentous macroalgae. It has been previously shown that *C. glomerata* is photosynthetically most active at high temperatures (20–30°C) and does not tolerate very low temperatures (Thomas, Collins, and Russell 1988). In addition, elevated nutrients increased the concentration of Chl in these green macroalga species in summer and autumn conditions. This is in agreement with the findings of Buapet et al. (2008) and Menendez, Herrera, and Comin (2002) who concluded that nutrient enrichment triggers an increase in Chl content in the green macroalgae *Ulva reticulata* and *Chaetomorpha linum* after 4 days of treatment. In their studies, however, the effect weakened after 7–10 days. Finally, reduced salinity conditions increased the pigment concentration of green filamentous macroalgae but the effect was reversed under enriched nutrient conditions. All of this suggests that context specificity is important for understanding how environmental variability modulates the pigment composition of macroalgae.

Nutrient enrichment did not have a straightforward relationship to charophyte *C. aspera* pigment concentration, whereas elevated nutrient load resulted in higher Chl-*a* and Chl-*b* concentration of the charophyte *C. horrida*. Temperature was the most influential environmental factor for both charophyte species. *C. aspera* showed highest Chl-*a* and Chl-*b* concentrations in summer under ambient temperature and *C. horrida* in autumn under elevated temperature. It can be assumed that photosynthetic characteristics of charophytes, as well as pigment concentrations, vary between different light conditions (Blindow et al. 2003), but this analysis was not included to our experiments.

5. Conclusions

We aimed at establishing relationship between measured pigments concentrations and spectral responses of different Baltic Sea macroalgae species. Our results showed that differences in Chl-*a* + *b* concentrations resulted in the consistent change of spectral reflectance for studied brown (*F. vesiculosus*) and green (*C. glomerata, U. intestinalis*) macroalgae species. The relationship was much weaker, however, in case of studied charophytes (*C. aspera, C. horrida*).

Different spectral indices have been more or less successful when relating reflectance spectra to pigment concentration. If spectral indices used in the current study predicted relatively well the patterns of Chl-*a* + *b* for most of the studied species, as well as across the five studied species, then the concentrations of Chl-*a* and Chl-*b* were more difficult to model. No good relationship was found between Car spectral index and the concentration of Car. The predictability of Car to Chl-*a* + *b* ratio varied largely among different macroalgae species. Nevertheless, we found a generic relationship between PRI and Car: Chl-*a* + *b* ratio across all studied macroalgae species.

The measurements of spectral reflectance of macroalgae were made using both a field portable spectrometer (Ramses, built by TriOS GmbH, Germany) and a laboratory hyperspectral imaging device (HySpex, built by Norsk Elektro Optikk, Norway). The HySpex imaging system yielded systematically better results in predicting pigment concentrations compared to the Ramses non-imaging instrument. In addition, by combining traditional assessment of pigment concentration with the use of the HySpex imaging device, it was possible to build models with a capability to predict the spatial patterns of pigment concentration in the Baltic Sea macroalgae.

In this study, key macroalgae species inhabiting the northeastern region of the Baltic Sea were exposed to different stresses related to future climate scenario of the region, and the concentrations of Chl-*a*, Chl-*b*, and Cars quantified to study macroalgae responses to environmental changes. Our results showed that different species were more (*C. glomerata, U. intestinalis, F. vesiculosus*) or less (*C. horrida, C. aspera*) influenced by environmental conditions; however, most of them showed higher pigment concentrations in case of elevated nutrients and temperature. Reduced salinity in general increased the Chl-*a* and Chl-*b* concentration under natural environmental conditions in summer and autumn but lowered the values under elevated temperature conditions in autumn.

Acknowledgements

This work was supported by the Estonian Research Council: [Grant number PUT1049].

Disclosure statement

No potential conflict of interest was reported by the authors.

Funding

This work was supported by the Eesti Teadusagentuur: [Grant Number PUT1049].

References

Alström-Rapaport, C., E. Leskinen, and P. Pamilo. 2010. "Seasonal Variation in the Mode of Reproduction of Ulva Intestinalis in a Brackish Water Environment." *Aquatic Botany* 93: 244–249. doi:10.1016/j.aquabot.2010.08.003.

Anderson, J., W. Chow, and D. Godchild. 1988. "Thylakoid Membrane Organisation in Sun/Shade Acclimation." *Australian Journal of Plant Physiology* 15: 11–26. doi:10.1071/PP9880011.

Andersson, A., H. E. M. Meier, M. Ripszam, O. Rowe, J. Wikner, P. Haglund, K. Eilola, et al. 2015. "Projected Future Climate Change and Baltic Sea Ecosystem Management." *Ambio* 44 (3): 345–356. doi:10.1007/s13280-015-0654-8.

Back, S., and A. Ruuskanen. 2000. "Distribution and Maksimum Growth Depth of Fucus Vesiculosus along the Gulf of Finland." *Marine Biological* 136: 303–307. doi:10.1007/s002270050688.

Bergsträsser, S., D. Fanourakis, S. Schmittgen, M. P. Cendrero-Mateo, M. Jansen, H. Scharr, and U. And Rascher. 2015. "HyperART: Non-Invasive Quantification of Leaf Traits Using Hyperspectral Absorption-Reflectance-Transmittance Imaging." *Plant Methods* 11:1.

Blackburn, G. A. 1998. "Spectral Indices for Estimating Photosynthetic Pigment Concentrations: A Test Using Senescent Tree Leaves." *International Journal of Remote Sensing* 19 (4): 657–675. doi:10.1080/014311698215919.

Blackburn, G. A. 2007. "Hyperspectral Remote Sensing of Plant Pigments." *Journal of Experimental Botany* 58 (4): 855–867. doi:10.1093/jxb/erl123.

Blindow, I. 2000. "Distribution of Charophytes along the Swedish Coast in Relation to Salinity and Eutrophication." *International Reviews Hydrobiol* 85: 707–717. doi:10.1002/1522-2632(200011)85:5/6<707::AID-IROH707>3.0.CO;2-W.

Blindow, I., J. Dietrich, N. Möllmann, and H. Schubert. 2003. "Growth, Photosynthesis and Fertility of *Chara Aspera* under Different Light and Salinity Conditions." *Aquatic Botany* 76: 213–234. doi:10.1016/S0304-3770(03)00053-6.

Bring, A., P. Rogberg, and G. Destouni. 2015. "Variability in Climate Change Simulations Affects Needed Long-Term Riverine Nutrient Reductions for the Baltic Sea." *Ambio* 44 (3): 381–391. doi:10.1007/s13280-015-0657-5.

Buapet, P., R. Hiranpan, R. J. Ritchie, and A. Prathep. 2008. "Effect of Nutrient Inputs on Growth, Chlorophyll, and Tissue Nutrient Concentration of Ulva Reticulata from a Tropical Habitat." *ScienceAsia* 34: 245–252. doi:10.2306/scienceasia1513-1874.2008.34.245.

Carter, G. A., and A. K. Knapp. 2001. "Leaf Optical Properties in Higher Plants: Linking Spectral Characteristics to Stress and Chlorophyll Concentration." *American Journal of Botany* 88 (4): 667–684. doi:10.2307/2657068.

Chapin, F. S. III, E. Rincon, and P. Huante. 1993. "Environmental Responses of Plants and Ecosystems as Predictors of the Impact of Global Change." *Journal Bioscience* 18 (4): 515–524. doi:10.1007/BF02703083.

Chappelle, E. W., M. S. Kim, and J. E. McMurtrey III. 1992. "Ratio Analysis of Reflectance Spectra (RARS): An Algorithm for the Remote Estimation of the Concentrations of Chlorophyll A, Chlorophyll B, and Carotenoids in Soybean Leaves." *Remote Sensing of Environment* 39: 239–247. doi:10.1016/0034-4257(92)90089-3.

Choo, K. S., J. Nilsson, M. Pedersen, and P. Snoeijs. 2005. "Photosynthesis, Carbon Uptake and Antioxidant Defence in Two Coexisting Filamentous Green Algae under Different Stress Condition." *Marine Ecology Progress Series* 292: 127–138. doi:10.3354/meps292127.

Croft, H., J. M. Chen, and Y. Zhang. 2014. "The Applicability of Empirical Vegetation Indices for Determining Leaf Chlorophyll Content over Different Leaf and Canopy Structures." *Ecological Complexity* 17: 119–130. doi:10.1016/j.ecocom.2013.11.005.

Gamon, J. A., J. Pen˜uelas, and C. B. Field. 1992. "A Narrow-Waveband Spectral Index that Tracks Diurnal Changes in Photosynthetic Efficiency." *Remote Sensing of Environment* 41: 35–44. doi:10.1016/0034-4257(92)90059-S.

Garbulsky, M. F., I. Filella, A. Verger, and J. Penuelas. 2014. "Photosynthetic Light Use Efficiency from Satellite Sensors: From Global to Mediterranean Vegetation." *Environmental and Experimental Botany* 103: 3–11. doi:10.1016/j.envexpbot.2013.10.009.

Gitelson, A. A., and M. Merzlyak. 1994. "Spectral Reflectance Changes Associated with Autumn Senescence of Aesculus Hippocastanum L. And Acer Platanoides L. Leaves: Spectral Features and Relation to Chlorophyll Estimation." *Journal Plant Physiological* 143: 286–292. doi:10.1016/S0176-1617(11)81633-0.

Gitelson, A. A., and M. N. Merzlyak. 1996. "Signature Analysis of Leaf Reflectance Spectra: Algorithm Development for Remote Sensing of Chlorophyll." *Journal Plant Physiological* 148: 494–500. doi:10.1016/S0176-1617(96)80284-7.

Gitelson, A. A., and M. N. Merzlyak. 1997. "Remote Estimation of Chlorophyll Content in Higher Plant Leaves." *International Journal of Remote Sensing* 18 (12): 2691–2697. doi:10.1080/014311697217558.

Graiff, A., D. Liesner, U. Karsten, and I. Bartsch. 2015. "Temperature Tolerance of Western Baltic Sea Fucus Vesiculosus - Growth, Photosynthesis and Survival." *Journal of Experimental Marine Biology and Ecology* 471: 8–16. doi:10.1016/j.jembe.2015.05.009.

International Panel on Climate Change (IPCC). 2013. *Climate Change 2013: The Physical Science Basis.* Cambridge, UK: Cambridge University Press.

Jones, H. G., and R. A. Vaughan. 2010. *Remote Sensing of Vegetation. Principles, Techniques and Applications.* Oxford, New York: Oxford University Press.

Jordan, C. F. 1969. "Derivation of Leaf Area Index from Quality of Light in the Forest Floor." *Ecology* 50: 663–666. doi:10.2307/1936256.

Joyce, K. E., and S. R. Phinn. 2003. "Hyperspectral Analysis of Chlorophyll Content and Photosynthetic Capacity of Coral Reef Substrates." *Limnology and Oceanography* 48: 489–496. doi:10.4319/lo.2003.48.1_part_2.0489.

Kautsky, H. 1992. "The Impact of Pulp-Mill Effluents on Phytobenthic Communities in the Baltic Sea." *Ambio* 21: 308–331.

Kirst, G. 1990. "Salinity Tolerance of Eukaryotic Marine Algae." *Annual Review of Plant Physiology and Plant Molecular Biology* 41: 21–53. doi:10.1146/annurev.pp.41.060190.000321.

Koch, M., G. Bowes, C. Ross, and X.-H. Zhang. 2013. "Climate Change and Ocean Acidification Effects on Seagrasses and Marine Macroalgae." *Glob Change Biologic* 19: 103–132. doi:10.1111/j.1365-2486.2012.02791.x.

Kotta, J., and T. Möller. 2014. "Linking Nutrient Loading, Local Abiotic Variables, Richness and Biomasses of Macrophytes, and Associated Invertebrate Species in the North-Eastern Baltic Sea." *Estonian Journal of Ecology* 63: 145–167. doi:10.3176/eco.2014.3.03.

Kotta, J., T. Möller, H. Orav-Kotta, and M. Pärnoja. 2014. "Realized Niche Width of a Brackish Water Submerged Aquatic Vegetation under Current Environmental Conditions and Projected Influences of Climate Change." *Marine Environmental Research* 102: 88–101. doi:10.1016/j.marenvres.2014.05.002.

Le Maire, G., C. Francois, and E. Dufrene. 2004. "Towards Universal Broad Leaf Chlorophyll Indices Using PROSPECT Simulated Database and Hyperspectral Reflectance Measurements." *Remote Sensing of Environment* 89: 1–28. doi:10.1016/j.rse.2003.09.004.

Leskinen, E., C. Alstrom-Rapaport, and P. Pamilo. 2004. "Phylogeographical Structure, Distribution and Genetic Variation of the Green Algae Ulva Intestinalis and U. Compressa (Chlorophyta) in the Baltic Sea Area." *Molecular Ecology* 13 (8): 2257–2265. doi:10.1111/j.1365-294X.2004.02219.x.

Lichtenthaler, H. K. 1987. "Chlorophylls and Carotenoids: Pigments of Photosynthetic Biomembranes." *Methods in Enzymology* 148: 350–382.

Meier, H. E. M., H. C. Andersson, B. Arheimer, T. Blenckner, B. Chubarenko, C. Donnelly, K. Eilola, et al. 2012. "Comparing Reconstructed past Variations and Future Projections of the Baltic Sea Ecosystem—First Results from Multi-Model Ensemble Simulations". *Environmental Research Letters* 7: 034005. doi:10.1088/1748-9326/7/3/034005.

Menendez, M., J. Herrera, and F. A. Comin. 2002. "Effect of Nitrogen and Phosphorus Supply on Growth, Chlorophyll Content and Tissue Composition of the Macroalga Chaetomorpha Linum (O.F. Müll.) Kütz in a Mediterranean Coastal Lagoon." *Journal of Marine Science* 66 (4): 355–364. doi:10.3989/scimar.2002.66n4355.

Merzlyak, M. N., A. A. Gitelson, O. B. Chivkunova, and V. Y. Rakitin. 1999. "Non Destructive Optical Detection of Pigment Changes during Leaf Senescence and Fruit Ripening." *Physiologia Plantarum* 106: 135–141. doi:10.1034/j.1399-3054.1999.106119.x.

Nicotra, A. B., M. Hofmann, K. Siebke, and M. C. Ball. 2003. "Spatial Patterning of Pigmentation in Evergreen Leaves in Response to Freezing Stress." *Plant, Cell and Environment* 26: 1893–1904. doi:10.1046/j.1365-3040.2003.01106.x.

Nygård, C. A., and M. J. Dring. 2008. "Influence of Salinity Temperature, Dissolved Inorganic Carbon and Nutrient Concentration on the Photosynthesis and Growth of Fucus Vesiculosus from the Baltic and Irish Seas." *European Journal of Phycology* 43 (3): 253–262.

Peñuelas, J., F. Baret, and I. Filella. 1995. "Semi-Empirical Indices to Assess Carotenoids/Chlorophyll a Ratio from Leaf Spectral Reflectance." *Photosynthetica* 31 (2): 221–230.

Peñuelas, J., and I. Filella. 1998. "Visible and Near-Infrared Reflectance Techniques for Diagnosing Plant Physiological Status." *Trends in Plant Science* 3: 151–156. doi:10.1016/S1360-1385(98) 01213-8.

Peñuelas, J., M. F. Garbulsky, and I. Filella. 2011. "Photochemical Reflectance Index (PRI) and Remote Sensing of Plant CO2uptake." *The New Phytologist* 191: 596–599. doi:10.1111/j.1469-8137.2011.03791.x.

Raven, J. A., and G. Samuelsson. 1988. "Ecophysiology of Fucus Vesiculosus L. Close to Its Northern Limit in the Gulf of Bothnia." *Botanica Marina* 31: 399–410. doi:10.1515/botm.1988.31.5.399.

Richardson, A. D., S. P. Duigan, and G. P. Berlyn. 2002. "An Evaluation of Noninvasive Methods to Estimate Foliar Chlorophyll Content." *New Phytologist* 153: 185–194. doi:10.1046/j.0028-646X.2001.00289.x.

Sims, D. A., and J. A. Gamon. 2002. "Relationship between Leaf Pigment Content and Spectral Reflectance across a Wide Range of Species, Leaf Structures and Developmental Stanges." *Remote Sensing of Environment* 81: 337–354.

Thomas, D. N., J. C. Collins, and G. Russell. 1988. "Interactive Effects of Temperature and Salinity upon Net Photosynthesis of Cladophora Glomerata (L.) Kütz. And C. Rupestris (L.) Kütz." *Botanica Marina* 31: 73–77. doi:10.1515/botm.1988.31.1.73.

Torn, K. (2008). "Distribution and ecology of charophytes in the Baltic Sea." Dissertationes Biologicae Universitatis Tartuensis., Tartu University Press, Tartu.

Torn, K., and G. Martin. 2004. "Environmental Factors Affecting the Distribution of Charophyte Species in Estonian Coastal Waters, Baltic Sea." *Proceedings Estonian Academic Sciences Biologic Ecology* 53 (4): 251–259.

Torres-Perez, J. L., L. S. Guild, R. A. Armstrong, J. Corredor, A. Zuluaga-Montero, and R. Polanco. 2015. "Relative Pigment Composition and Remote Sensing Reflectance of Caribbean Shallow-Water Corals." *PLoS ONE* 10 (11): e0143709. doi:10.1371/journal.pone.0143709.

Ustin, S. L., G. P. Asner, J. A. Gamon, K. F. Huemmrich, S. Jacquemoud, M. Schaepman, and P. Zarco-Tejada. 2006. *Retrieval of Quantitative and Qualitative Information about Plant Pigment Systems from High Resolution Spectroscopy*. New York, NY: IEEE.

Ustin, S. L., D. A. Roberts, J. A. Gamon, G. P. Asner, and R. O. Green. 2004. "Using Imaging Spectroscopy to Study Ecosystem Processes and Properties." *Bioscience* 54 (6): 523–534. doi:10.1641/0006-3568(2004)054[0523:UISTSE]2.0.CO;2.

Vogelmann, J. E., B. N. Rock, and D. M. Moss. 1993. "Red Edge Spectral Measurements from Sugar Maple Leaves." *International Journal of Remote Sensing* 14 (8): 1563–1575. doi:10.1080/01431169308953986.

Assessment of PlanetScope images for benthic habitat and seagrass species mapping in a complex optically shallow water environment

Pramaditya Wicaksono and Wahyu Lazuardi

ABSTRACT

This paper presents the first assessment of PlanetScope image for benthic habitat and seagrass species mapping in optically shallow water. PlanetScope image is equipped with ideal resolutions for benthic habitat and seagrass mapping including high-spatial resolution (3 m), high radiometric resolution (12-bit), sufficient water penetration bands (Visible-Near-infrared) and very high temporal resolution (almost daily), which distinguishes it from other high spatial resolution images. It is necessary to assess the accuracy of this ideal system in a real-world benthic habitat and seagrass species mapping application. The optically shallow water of Karimunjawa Islands was selected as the study area. Two PlanetScope images acquired on 17 May 2017 and 15 August 2017 were tested as a control for the consistency of PlanetScope image accuracy. Several treatments were applied to both PlanetScope images including atmospheric correction, sunglint correction, Principle Component Analysis (PCA), Minimum Noise Fraction (MNF) and Linear Spectral Unmixing (LSU). Per-pixel classification algorithms (including Maximum Likelihood – ML, Support Vector Machine – SVM, and Classification Tree Analysis – CTA) and Object-based Image Analysis (OBIA) were used to perform benthic habitat and seagrass species classification. Spectra-based classifications were also applied to classify seagrass species using seagrass species spectra as input endmember. The results indicated that PlanetScope images produced 47.13–50.00% overall accuracy (OA) for benthic habitat mapping consist of five classes (coral reefs, macroalgae, seagrass, bare substratum, dead coral) and 74.03–74.31% OA for seagrass species mapping consist of five seagrass species classes. The accuracy of PlanetScope images for benthic habitat and seagrass species mapping was comparable to other high spatial resolution images. The performance of PlanetScope images was also consistent, shown by the similar accuracy obtained from May and August image. The concern regarding PlanetScope image was the low Signal-to-Noise Ratio (SNR) over homogeneous areas such as optically deep water, which led to the failure of performing sunglint correction and obtaining higher accuracy. To conclude, with the very high temporal resolution, PlanetScope image is promising for monitoring the dynamics and changes of benthic habitat and seagrass species composition, and rapid assessment of extreme events impacts, especially in coastal areas with limited accessibility.

1. Introduction

PlanetScope, as the newest high spatial resolution satellite imaging, is capable of recording earth's surface area of 150 million km^2 per day. According to Planet (2017), each PlanetScope satellite is a CubeSat 3U form factor (10 × 10 × 30 cm). The complete PlanetScope constellation is approximately 120 satellites. They will be able to produce satellite images of Earth surface everyday. PlanetScope satellite has a Sun-Synchronous orbit and an International Space Station (ISS) orbit with four visible bands (Red, Green, Blue, Near-infrared (NIR)), 3 m spatial resolution, 12-bit radiometric resolution, as well as having higher temporal resolution than other governmental or commercial satellites (approximately 1 day) (Planet 2017). With these characteristics, utilizing PlanetScope images for mapping applications has many advantages, especially in providing real-time information for understanding impacts in case of extreme weather and disaster event occurred. Based on its spatial resolution (3 m), it is capable of generating detailed information for large area, and from its very high temporal resolution, it is possible to obtain an image with the same recording date as the field survey and high temporal monitoring effort.

Despite the specifications, assessment of PlanetScope image capability to map natural resources is still limited, primarily related to mapping underwater coastal resources such as benthic habitat including coral reefs and seagrass. Benthic habitat mapping is always a challenging application due to the limitation of remote-sensing sensors and benthic habitat environmental complexities (Hedley et al. 2012). Seagrass species mapping using remote-sensing is also a challenging task because reflectance recorded by remote-sensing is not only from the seagrass but also from the sunglint, water column, and atmospheric disturbance. Hence, the variation of seagrass reflectance is not only due to the variation of seagrass object in the field, i.e., species, biomass, but also due to the variation of atmospheric condition, sunglint and water column condition. It was shown by the variety of results and accuracies of previous works (Phinn et al. 2008; Lyons, Phinn, and Roelfsema 2011; Roelfsema et al. 2014; Hedley et al. 2017). There are various image processing methods applied to perform benthic habitat mapping including the application of image radiometric corrections (atmospheric, sunglint, water-column), image transformation (Principle Component Analysis – PCA, Minimum Noise Fraction – MNF), and image classifications (Object-based Image Analysis – OBIA, per-pixel classification algorithms) (Green et al. 2000; Lyons, Phinn, and Roelfsema 2011; Phinn, Roelfsema, and Mumby 2012; Goodman, Purkis, and Phinn 2013). The variations of these image processing approach to perform benthic habitat and seagrass species mapping lead to the variation of the obtained accuracy. For instance, Phinn et al. (2008) reported a very low accuracy of seagrass species mapping (7 classes, <30%) when using per-pixel classification, meanwhile, Roelfsema et al. (2014) produced much higher accuracy of seagrass species map using OBIA (5 class, >60%) although the accuracy difference is partially affected by the difference in the seagrass species classification scheme. The variety of classification scheme used by different researchers also determines the accuracy of the result. Also, seagrass habitat complexity of the study area and the suitability of remote-sensing image used to capture the variation of benthic habitat and seagrass species *in situ* also control the accuracy. Thus, it is necessary to seek understanding whether PlanetScope image can follow the success of its predecessors such as WorldView-2, IKONOS, and Quickbird (Phinn, Roelfsema, and Mumby 2012; Roelfsema et al. 2014).

PlanetScope image needs to be evaluated for benthic habitat and seagrass species mapping because its specification may address several issues encountered when using the existing satellite images, i.e., the delay between image acquisition and field survey, high temporal resolution images for monitoring routines and for rapid assessment during extreme events. The successful use of PlanetScope image for benthic habitat and seagrass species mapping will have a positive impact, since its high temporal resolution will be very beneficial for monitoring the dynamics and changes of benthic habitat and seagrass, and rapid assessment of extreme events, especially in coastal areas with limited accessibility. This work is part of the Planet Education and Research program (PlanetTeam 2017).

This paper presents the first assessment of PlanetScope multi-temporal images for mapping benthic habitat and seagrass species in complex optically shallow water of Karimunjawa Islands, Indonesia. Benthic habitats provide various ecosystem services which are ecologically and economically important (Kritzer et al. 2016). Benthic habitat map is strongly required for rapid assessment of habitat health and responses to stress, conservation and management of coral reefs ecosystem, and habitat types inventory and their dynamics (Zhang et al. 2013). Management and monitoring of seagrass species biodiversity require spatial information on a detailed-scale (Bell, Fonseca, and Stafford 2006; Kenworthy et al. 2006). Seagrass species have an important role in the preservation of the biodiversity of coastal ecosystems (Larkum, Orth, and Duarte 2006), as they provide various ecosystem services as reviewed by Duffy (2006) and Nordlund et al. (2016). They found out that *Enhalus acoroides* beds provide habitat for fish and marine invertebrate, nursery ground, food for human, carbon sequestration, sediment stabilization, and coastal protection. Meanwhile, *Thalassia hemprichii* and *Cymodocea rotundata* can also be used as compost fertilizer in addition to providing the function similar to *Enhalus acoroides*. Therefore, seagrass species mapping is necessary to map the unique ecosystem services provided by different seagrass species. As a consequence, mapping and monitoring of benthic habitat and seagrass biodiversity on a detailed-scale is a vital component in the management and monitoring of coastal area (Green et al. 2000). Understanding seagrass species spatial distribution can also be related to the appearance of specific marine biotas such as sea turtle and dugong. One of the best options to address these issues is to utilize multispectral remote-sensing system that provides an effective and efficient data source to map coastal resources, especially seagrass species biodiversity. The utilization of remote-sensing images to map seagrass species is limited to high spatial resolution and limited to optically shallow coastal water (Mumby and Green 2000; McKenzie, Finkbeiner, and Kirkman 2001; Dekker et al. 2006; Phinn et al. 2008; Hossain et al. 2015). Furthermore, remote-sensing approach can be used to assess temporal changes of seagrass species composition, which can be used as a basis for monitoring the dynamics of seagrass coverage and calculate the richness and biodiversity of seagrass species.

2. Study area and data

2.1. *Study area*

We selected several islands in Karimunjawa Islands as the study area. These islands are Karimunjawa Island, Kemujan Island, Menjangan Besar Island, Menjangan Kecil Island,

Gosong Island, and Cilik Island (Figure 1). The condition of benthic habitat in Karimunjawa Islands is relatively stable and protected since it is under the authority of Karimunjawa National Park since 1999. Different types of reef morphology are present in Karimunjawa Islands. Reef Flat, Reef Cut, Fore Reef, Bank/Shelf, and Escarpment are commonly found in Karimunjawa Island, Menjangan Besar Island, Menjangan Besar Island, Gosong Island and Cilik Island. Kemujan Island has all the aforementioned types of reef morphology with the addition of shallow Lagoon and Back Reef. Karimunjawa Islands also shelter a high ecological-diversity of benthic habitat composition (Wicaksono 2016). Based on our bathymetry model, the sunlight may penetrate water up to the depth of 17.6 m (Wicaksono 2010), however, the optically shallow water

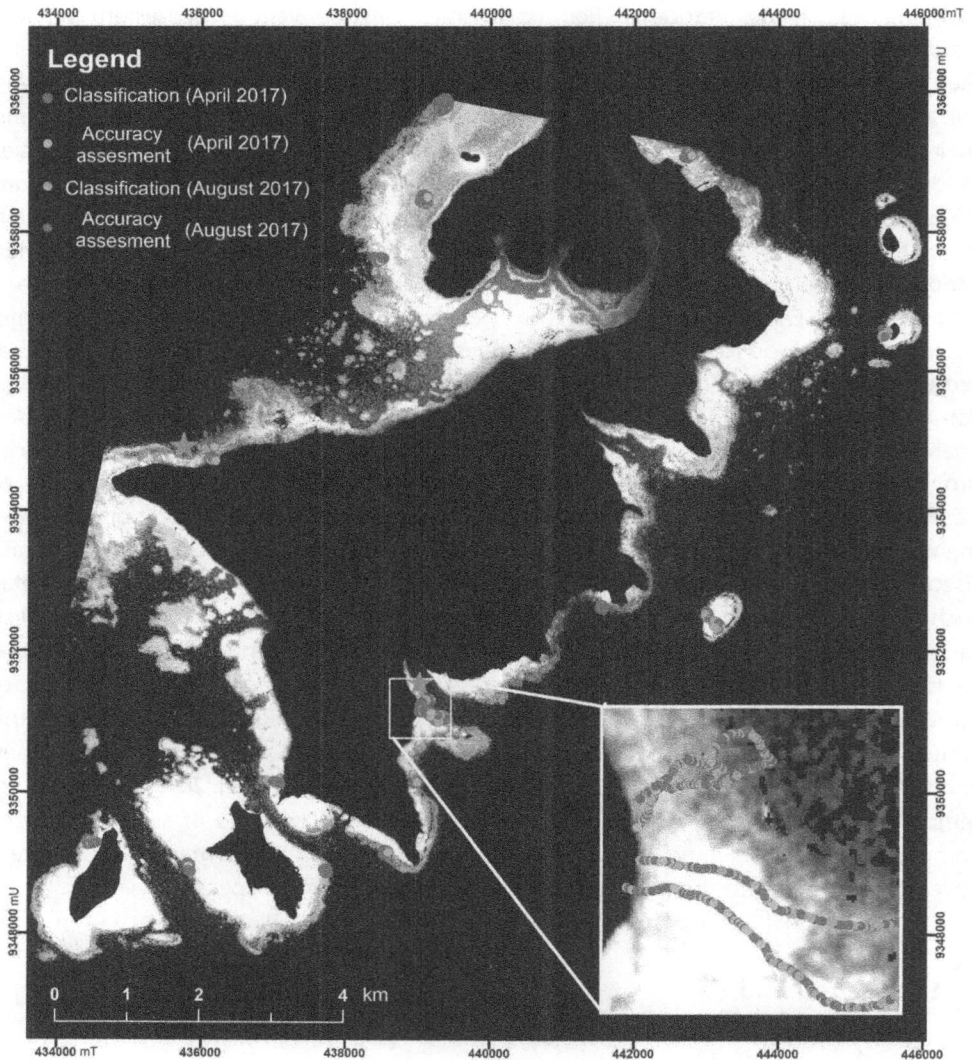

Figure 1. Photo-transect samples distribution from the field survey conducted on 8–13 April 2017 and 11–16 August 2017 (overlaid on PlanetScope image recorded on 15 August 2017).

where the seafloor is still effectively visible from the remote-sensing image is only up to the depth of 11.6 m (Wicaksono 2015). Therefore, we only effectively mapped optically shallow water up to the depth of 11.6 m.

2.2. *Image data*

Two PlanetScope images at the 3B level were used in this research (Table 1). We obtained the images for free as this work is part of Planet Education and Research program (PlanetTeam 2017). PlanetScope 3B level image is an Ortho Scene Product, which is orthorectified, and the pixel value is scaled to Top-of-Atmosphere (TOA) radiance (at-sensor). This product has scene-based framing and projected to a Universal Transverse Mercator (UTM) cartographic projection (Planet 2017). The use of two PlanetScope images was necessary as a control for radiometric quality variations of PlanetScope images so that the conclusion regarding the performance of PlanetScope images is not only by chance (which is may be due to the radiometric quality of a single image) but justified by more than one image. Presently, Planet made the surface reflectance (SR) product available (Collison and Wilson 2017), and thus, the orthorectified PlanetScope images no longer need atmospheric correction. However, by the time we performed this research, the SR image product is not yet available. The date of image acquisition for both PlanetScope images was 17 May 2017 and 15 August 2017 respectively. These images were the best images recorded closest to the date of field survey. Image acquired exactly at the date of field survey, especially for the survey conducted in April 2017, was not usable due to severe cloud issue covering the benthic habitats.

2.3. *Field data*

The field surveys were conducted in 8–13 April 2017 and 11–16 August 2017 (Figure 1). Benthic habitat and seagrass species data were collected in the field using photo-transect method to obtain underwater photos of benthic habitats and seagrass species composition (Roelfsema and Phinn 2009). In short, the surveyor swims and took underwater photo in constant interval every two paddles. The Garmin 78s Global Positioning System (GPS) carried by the surveyor was set in tracking mode and recorded the coordinate of the surveyor every three seconds. Afterward, each photo was given a UTM coordinate based on the time-match between photo and GPS recording. The locations of the photo-transect survey were selected based on the variation and representativeness of benthic habitats and seagrass species as observed from the image and based on our experience

Table 1. Specification of the two PlanetScope images including the reflectance coefficient used to convert the Digital Number (DN) into Top-of-Atmosphere (TOA) reflectance. The PlanetScope images were already orthorectified and projected to a cartographic projection, sensor corrections applied, and the pixel value is scaled to TOA radiance (at sensor). Since we obtained the image via GUI not API, the PlanetScope images were obtained at DN level.

Band	Band range (nm)	Reflectance Coefficient (17 May 2017)	Reflectance Coefficient (15 August 2017)
Blue	455 – 515	$2.28474870169 \times 10^{-5}$	$2.24233633246 \times 10^{-5}$
Green	500 – 590	$2.42001840134 \times 10^{-5}$	$2.37002630646 \times 10^{-5}$
Red	590 – 670	$2.69511032799 \times 10^{-5}$	$2.64415939736 \times 10^{-5}$
Near-Infrared	780 – 860	$4.05633512782 \times 10^{-5}$	$3.95124472761 \times 10^{-5}$

conducting research in this area since 2008. These variations include the combination of reefs geomorphic class such as reef flat, back reef, fore reef, lagoon, major benthic habitat variation as seen visually from PlanetScope image, and also the landscape of coastal area such the facing direction of the coastal area, beach materials, land cover, incoming currents. The locations of transect for the photo-transect survey for April and August survey were not similar. The purpose is to test whether the performance of PlanetScope image for benthic habitat mapping is consistent across different sample variations. We obtained 1,614 photo-transect samples from April Survey and 1,368 photo-transect samples from August survey (Table 2).

The photo-transect samples were analysed and interpreted using Coral Point Count Excel (CPCE) program (Kohler and Gill 2006). CPCE is an extension program for Microsoft Excel to interpret benthic habitats from underwater photos collected from the photo-transect survey. For benthic habitat mapping, the samples were categorized into five classes: coral reefs, seagrass, macroalgae, bare substratum, and dead coral. The labelling process of each photo was conducted based on the most dominant benthic class in each photo. Therefore, we have a training area collected in April and August used for the classification of May and August image, respectively. The photo-transect samples for each benthic habitat class were randomly categorized as the training area for the multispectral classification and for accuracy assessment of classification results.

Seagrass samples collected in August were used as training areas to classify both May and August PlanetScope image. This was preferred because the spatial distribution of seagrass species is more limited compared to coral reefs, macroalgae and bare substratum. Thus, particular species only present at a particular area, and as a consequence, it is not feasible to find samples of particular seagrass species in another area. There are no

Table 2. Sample number per class for classification training area and accuracy assessment samples. The class descriptor for each seagrass species class is also given. Ea = *Enhalus acoroides*, Th = *Thalassia hemprichii*, Cr = *Cymodocea rotundata*, Cs = *Cymodocea serrulata*, Hu = *Halodule uninervis*, Ho = *Halophila ovalis*.

Class	Training Area		Accuracy assessment		Seagrass class descriptor
	May	August	May	August	
Coral reefs	302	252	191	210	
Dead Coral	274	108	239	97	
Macroalgae	106	116	87	93	
Bare Substratum	228	185	187	175	
Seagrass	214 (67 samples collected in August)		362 (65 samples collected in August)		
Ea		16		30	Ea-dominated with small coverage of *Th* and or *Cr*. This include epiphyte-covered *Ea*
EaThCr		12		22	Ea mixed with *Th* and or *Cr*
Th		7		12	Th-dominated bed with small coverage of *Cr*. *Cs* is also included in this class due to its similar morphology with *Th* and its very-localized spatial distribution.
ThCr		132		227	Th mixed with *Cr* with relatively similar proportion
CrHu		47		71	Cr-dominated bed with small coverage of *Hu*. *Ho* is also included in this class due to the very low coverage of *Ho* and it is mostly found associated with *Hu*. The distribution of *Hu* and *Ho* is also very localized
Total	**1124**	**875**	**704**	**640**	

significant changes in the composition and spatial distribution of seagrass species in April and August, and thus applying samples collected in August to train and validate May image is acceptable. In addition, we also added seagrass species samples collected in 2012–2013 since the spatial distribution of seagrass species is consistent between these time ranges. We confirmed this based on our survey and working experience in this island since 2008 (Wicaksono and Hafizt 2013, Wicaksono 2015). The samples were categorized based on the variation of seagrass species *in situ*. The commonly found seagrass species found during the survey were *Enhalus acoroides* (Ea), *Thalassia hemprichii* (Th), *Cymodocea rotundata* (Cr), *Cymodocea serrulata* (Cs), *Halodule uninervis* (Hu), and *Halophila ovalis* (Ho). At some locations, these species do not produce homogenous meadow instead they are mixing with other species such as *Ea* and *Th* or *Cr* (*EaThCr*), *Th* and *Cr* (*ThCr*), and *Cr* and *Hu* (*CrHu*). There is a mixing between *Hu* and *Ho* but very localized (in water of northern coast of Ujung Gelam) (see blue star in Figure 1), and the extent is so small with coverage mostly below 50%. The reflectance of this class resembles carbonate sand reflectance instead of seagrass. *Cs* only presents in coastal water of Legon Lele (see orange star in Figure 1) and in some eastern parts of Karimunjawa Island.

3. Methods

3.1. *Seagrass species spectra*

The reflectance spectra of seagrass species were used as endmember for Spectral Angle Mapper (SAM), Spectral Information Divergence (SID), and Linear Spectral Unmixing (LSU), and were adapted from Wicaksono et al. (2017). Among the species reported in Wicaksono et al. (2017), we only selected *Ea, Th, Cr, Cs,* and *Hu* as the most dominant seagrass species in the study area, and mostly found in the photo-transect samples that we collected during the survey. To match the classification scheme used in classification processes, we combined the reflectance spectra accordingly. The reflectance spectra for *CrHu* class was obtained by averaging the reflectance spectra of *Cr* and *Hu*. Similar approach was applied to obtain the reflectance spectra for *EaThCr* and *ThCr* class. Afterwards, these combined reflectance spectra of seagrass species were resampled into PlanetScope image spectral resolution by averaging the species reflectance spectra according to each PlanetScope band range. We used this approach since we do not have the pre-defined filter function of PlanetScope Image. The resampled reflectance spectra were used as endmember for classifying seagrass species using SAM and SID algorithm. The use of resample reflectance spectra as endmember for LSU was to obtain image fraction of seagrass species. The fractions of seagrass species were used as input for classifying seagrass species using Maximum Likelihood (ML), Support Vector Machine (SVM), and Classification Tree Analysis (CTA). The resampled seagrass species reflectance spectra are shown in Figure 2.

3.2. *Image corrections*

3.2.1. *Atmospheric correction*
Both PlanetScope images were converted from DN to TOA reflectance by multiplying the DN value with the reflectance conversion coefficients, as described in the

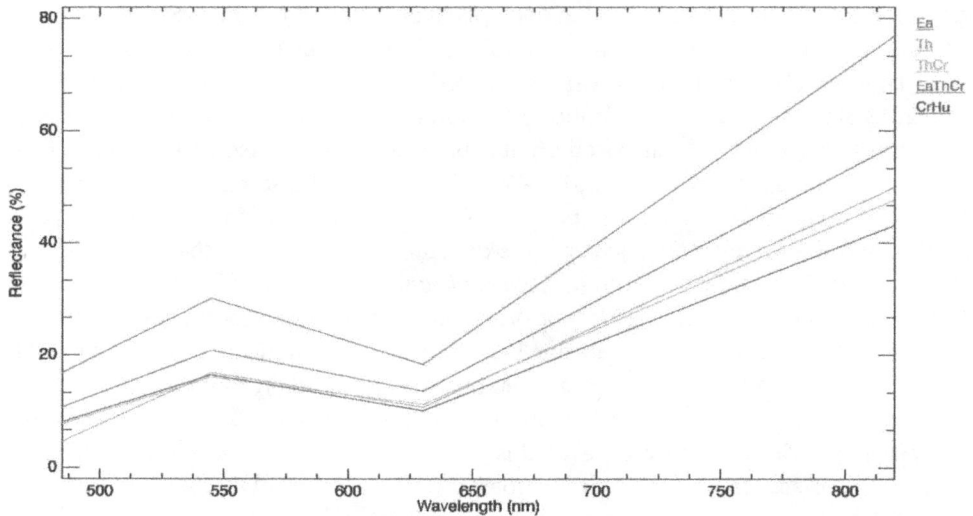

Figure 2. Seagrass species reflectance spectra used as endmember for Spectral Angle Mapper (SAM), Spectral Information Divergence (SID), and Linear Spectral Unmixing (LSU) input resampled to PlanetScope spectral resolution. For the original reflectance spectra of each species refer to Wicaksono et al. (2017).

corresponding image metadata (Table 1). Afterwards, simple atmospheric correction the Dark-Object Subtraction (DOS) method was applied, by using clear water free from sunglint to obtain Bottom-of-Atmosphere (BOA) reflectance image. The selection of dark-target for DOS followed the results from Wicaksono and Hafizt (2018), which was the optically deep water pixel free from sunglint, since it was assumed to be the object with the highest absorption.

3.2.2. *Sunglint correction*

A somewhat complicated issue was encountered during sunglint correction of both images. Both images have different sunglint intensities, where April image has higher sunglint intensity than August image. However, when attempted to remove the sunglint, we encountered difficulties in obtaining suitable empirical model between visible bands and NIR band. Despite the methods (Kay, Hedley, and Lavender 2009), sunglint correction approach requires a good correlation between visible and NIR bands since NIR band is required to remove sunglint in visible bands, and the correlation must present between them (Hedley, Harborne, and Mumby 2005).

The waterbody pixels of PlanetScope image in both images were noisy, which is the first indication of low Signal-to-Noise Ratio (SNR). We also suspected that there was a saturation issue on NIR band over optically deep water. As a consequence, it was not possible to obtain a strong correlation between visible bands and NIR band. The correlation between visible bands was mostly strong, although there were times when the correlation was low, and the nature of the relationship was inconsistent. The correlation between visible bands and NIR band was always weak, almost no correlation at all, and inconsistent. We concluded this after performing correlation analysis using pixel samples of different sunglint intensities across the scene for both images. There were times when there is a negative correlation between blue, green, and red band interchangeably. In addition, there was no significant correlation between visible bands and NIR band, and most of the times the correlation was negative. The negative correlation between bands for water pixels is not expected due to the similarity of the index of refraction between the wavelengths of those bands (Mobley 1994). This contrasts with the effort of removing sunglint in IKONOS, WorldView-2 or Quickbird image (Hedley, Harborne, and Mumby 2005; Kay, Hedley, and Lavender 2009; Wicaksono 2016). The positive correlation between bands for these images was strong and consistent. Hence, the empirical model for removing the sunglint was possible. PlanetScope image resolutions are comparable with these three images, and thus should perform similarly. We tried to manually-eliminate the outliers from the model, but despite the increase in the correlation coefficient of the model, the resulting sunglint-free image was noisy, and there was no benefit of sunglint removal. Due to these circumstances, we preferred not to perform sunglint correction as it adversely impacts the radiometric quality of PlanetScope image. This is an issue that should become a concern in the future development of Planet image constellation.

3.3. *Image transformation*

Principle Component Analysis (PCA) and Minimum Noise Fraction (MNF) transformation were applied to the atmospherically corrected PlanetScope images (visible bands only, excluding NIR band due to its limited water penetration capability) due to the spectral band limitation of PlanetScope image. Furthermore, the PlanetScope bands are highly correlated and the information is redundant between bands. Thus, to improve the number of the band with unique information, we applied PCA to produce unique and uncorrelated bands to improve benthic habitat separability (Wicaksono 2016). Wicaksono (2016) assessed the use of PCA and Independent Component Analysis (ICA) transformation for benthic habitat mapping. The application of PCA in the WorldView-2 multispectral image has successfully improved the accuracy of benthic habitat mapping effectively at various levels of benthic habitat classification schemes. The highest classification accuracy at each level of the classification scheme was obtained through PC band. In addition, noise was also high in PlanetScope image (low SNR), which indicated by the high standard deviation of a homogenous object such as the optically deep water. The MNF transformation improved the quality of data input and provided better results than atmospherically corrected image (Ayoobi and Tangestani 2017). In the context of benthic habitat mapping, Zhang et al. (2013) applied MNF in AVIRIS hyperspectral image to suppress the redundancy of spectral information between bands and managed to improve the accuracy of benthic habitat mapping. Applying MNF may help separate the data from the noise in PlanetScope image. In this research, we included three PC bands and three MNF bands in the classification process because relevant information regarding benthic habitat and seagrass species may be located in the later band and not always in the PC or MNF band with high eigenvalue. There might be a chance that the useful information is not included in the classification when selecting only PC and MNF bands with high eigenvalue.

3.4. *Benthic habitat mapping*

The input bands used for benthic habitat classification using per-pixel classification and OBIA were BOA reflectance bands, PC bands, and MNF bands. We performed the classification using BOA reflectance bands, PC bands, MNF bands, and seagrass species image fraction individually because we wanted to analyse how different treatments will affect the accuracy of PlanetScope image (Table 3).

3.4.1. *Per-pixel classification*

Benthic habitat classification was performed using Maximum Likelihood (ML), Support Vector Machine (SVM), and Classification Tree Analysis (CTA) algorithms. The three algorithms were selected to test the capabilities of PlanetScope image because each algorithm has different characteristics in the pixel-clustering process. ML is the most commonly used classification algorithm in benthic habitat mapping (Green et al. 2000; Casal et al. 2011; Goodman, Purkis, and Phinn 2013). ML algorithm classifies pixels based on the probability density function derived from the statistics of the training areas (Richards 2013). Pixels are classified into the class where they have the highest probability to belong to this particular class. SVM is a powerful machine-learning technique for image classification that has the capability of creating boundaries called the hyperplane in the multi-dimensional feature space to separate and classify each pixel into classes (Vapnik 1995; Huang, Davis, and Townshend 2002). SVM can provide better accuracy of benthic habitat classification than ML, although it is not as widely used as ML (Eugenio, Marcello, and Martin 2015; Zhang 2015; Wahidin et al. 2015). CTA is a machine-learning classification method that utilizes decision tree rules as the main algorithm, so it is effective to classify complex objects such as benthic habitat and seagrass species. The training area for classification process is benthic habitat field data obtained from the interpretation of underwater photo collected from the photo-transect survey.

3.4.2. *Object-Based Image Analysis (OBIA)*

OBIA applied in this research consists of two main processes, namely image segmentation and multispectral classification. Image segmentation was done by taking into

Table 3. Summary of multispectral classification for benthic habitat and seagrass species mapping using PlanetScope images. OBIA: Object-based Image Analysis, ML: Maximum Likelihood, SVM: Support Vector Machine, CTA: Classification Tree Analysis, SAM: Spectral Angle Mapper, SID: Spectral Information Divergence, LSU: Linear Spectral Unmixing, BOA: Bottom-of-Atmosphere, PC: Principle Component, MNF: Minimum Noise Fraction.

	Classification type	Classification algorithm	Input band
Benthic habitat mapping	Per-pixel	ML, SVM, CTA	BOA reflectance bands, PC bands, MNF bands
	OBIA	Image segmentation with ML, SVM, CTA	BOA reflectance bands, PC bands, MNF bands
Seagrass species	Per-pixel	ML, SVM, CTA	BOA reflectance bands, PC bands, MNF bands, fraction of seagrass species from LSU
	OBIA	Image segmentation with ML, SVM, CTA	BOA reflectance bands, PC bands, MNF bands, fraction of seagrass species from LSU
	Spectra-based	SAM, SID	BOA reflectance bands

account the similarity of the character of an object or pixel and making it a specific object segment according to the threshold scale parameter setting being used (Benz et al. 2004). The process of image segmentation was performed using IDRISI Selva using parameters such as moving-window size, similarity tolerance, mean, and variance. The resulting-segments were classified into benthic habitat using training area collected in the field. The multispectral classification algorithms used were ML, SVM, and CTA.

3.5. *Seagrass species mapping*

Seagrass species mapping was performed on seagrass pixels using seagrass mask obtained from the benthic habitat map. There were several benthic habitat classification results, obtained from SVM, ML, CTA, and OBIA. Each produced different seagrass spatial distribution, user's accuracy, and producer's accuracy. We selected the classification result that produced the highest seagrass user's and producer's accuracy and used it as a seagrass mask. The inputs for seagrass species mapping using per-pixel algorithms (ML, SVM, CTA), and OBIA (image segmentation with ML, SVM, CTA), were BOA reflectance bands, PC bands, MNF bands, and fraction of endmember of seagrass species obtain from Linear Spectral Unmixing (LSU) analysis. LSU analysis (Van der Meer and de Jong 2001) was applied to the BOA reflectance image to obtain the fraction of seagrass species on each seagrass pixel. The seagrass species endmember for obtaining seagrass species fraction from LSU was adapted from the Wicaksono et al. (2017) and resampled to PlanetScope image spectral resolution. The training area from photo-transect was used for seagrass species classification.

For spectra-based classification, Spectral Angle Mapper (SAM) and SID (Spectral Information Divergence) were applied using BOA reflectance bands only, as the seagrass species reflectance spectra used as the input endmember only match the reflectance of BOA reflectance bands. SAM has been previously used for benthic habitat mapping with various degrees of accuracy (Kutser, Miller, and Jupp 2006; Casal et al. 2011; Wicaksono et al. 2017). The use of SID was experimental since there is no report of previous work using SID for benthic habitat mapping. Summary of classification scenarios used in this research is provided in Table 3.

3.6. *Accuracy assessment*

Accuracy assessment of benthic habitat and seagrass species mapping was performed using confusion matrix analysis (Congalton and Green 2008). This analysis produced overall classification accuracy (OA), Kappa coefficient, and user's accuracy (UA) and producer's accuracy (PA) for each classified class. Information on UA and PA are necessary to assess the individual accuracy of either benthic habitat and seagrass species class, which is important to determine which class has higher or lower accuracy compared to others. The UA and PA were also used to assess whether a class extent is overestimated or underestimated. In addition, confusion matrix was used to analyse the misclassification rate between classes, which is necessary to determine which benthic habitat or seagrass classes are easily misclassified. The Kappa value was evaluated based on the strength of agreement described in Landis and Koch (1977): Poor (<0.00), Slight (0.00–0.20), Fair (0.21–0.40), Moderate (0.41–0.60), Substantial (0.61–0.80), Almost Perfect (0.80–1.00). Afterwards, the accuracy of PlanetScope image for benthic habitat and seagrass species mapping was compared with the accuracy of previous works that utilized other high spatial resolution image with almost similar characteristics. This was necessary to see how the performance of PlanetScope image compared to the existing high spatial resolution images for benthic habitat and seagrass species mapping.

4. Results

4.1. *Benthic habitat mapping*

4.1.1. *Benthic habitat mapping of 17 May 2017 image*

PlanetScope image produced benthic habitat map with a maximum OA of 47.13% (Kappa 0.33), obtained from CTA classification using BOA reflectance bands (Table 4). CTA produced the best accuracy for BOA reflectance bands, PC bands, and MNF bands. CTA delivered both the highest OA and Kappa agreement between classes. Nevertheless, the accuracy difference between ML, SVM, and CTA for mapping benthic habitat at this level of complexity was within 15% differences (33.05–47.13%). The OA of SVM was second best after CTA, but the agreement for each classified-benthic class with *in situ* data was worse than ML. Not all benthic habitat class can be classified by SVM algorithm, for instance, SVM failed to classify macroalgae despite the accuracy.

Figure 3 presents the UA and PA for the most accurate result of each input bands. The accuracy pattern from CTA results was consistent across different classification inputs. The spatial distribution and extent of coral reefs, bare substratum, and dead coral class are underestimated (UA is higher than PA), while seagrass and macroalgae class are overestimated (PA is higher than UA).

The maximum accuracy from OBIA was 46.84% (Kappa 0.33) using BOA reflectance bands. Map produced from OBIA of MNF bands produced a slightly lower OA, which was 46.22% (Kappa 0.32). OBIA of BOA reflectance bands and MNF bands produced a similar pattern of the UA and PA for each class (Figure 4). The spatial distribution of coral reefs, bare substratum, and dead coral was slightly underestimated, as indicated by the lower PA compared to UA. Macroalgae class was the most challenging class to map, and the extent was overestimated. Only about 37.93% of macroalgae in the field were correctly classified, with less than 20% chance of pixels classified as macroalgae were actually macroalgae in the field. More than 80% chance of pixels classified as macroalgae were not macroalgae in the field. This was similar to the seagrass class where the PA was almost 20% higher than the UA, indicating that many pixels classified as seagrass were not seagrass in the field. However, this overestimation of seagrass extent was preferred since seagrass species *in situ* samples have more chance to be included within the seagrass mask. The UA of seagrass class from OBIA of MNF bands was higher than from

Table 4. Accuracy assessment of benthic habitat mapping of PlanetScope image acquired on 17 May 2017. Values in bold are the highest overall accuracy (OA) for each input.

Algorithm	BOA reflectance bands		PC bands		MNF bands	
	OA (%)	Kappa	OA (%)	Kappa	OA (%)	Kappa
ML	33.05	0.20	42.48	0.27	42.48	0.27
SVM	45.76	0.26	40.33	0.20	40.00	0.19
CTA	47.13	0.33	46.09	0.32	43.61	0.29

(a)

	Coral reefs	Seagrass	Macro algae	Bare substratum	Dead coral
▪ User's accuracy (%)	51.35	51.00	23.84	56.99	54.32
▪ Producer's accuracy (%)	39.79	79.69	47.13	56.68	36.82

(b)

	Coral reefs	Seagrass	Macro algae	Bare substratum	Dead coral
▪ User's accuracy (%)	50.94	47.83	21.31	61.29	51.35
▪ Producer's accuracy (%)	42.41	68.75	44.83	60.96	31.80

(c)

	Coral reefs	Seagrass	Macro algae	Bare substratum	Dead coral
▪ User's accuracy (%)	46.05	55.06	18.44	57.45	46.87
▪ Producer's accuracy (%)	36.65	76.56	37.93	57.75	31.38

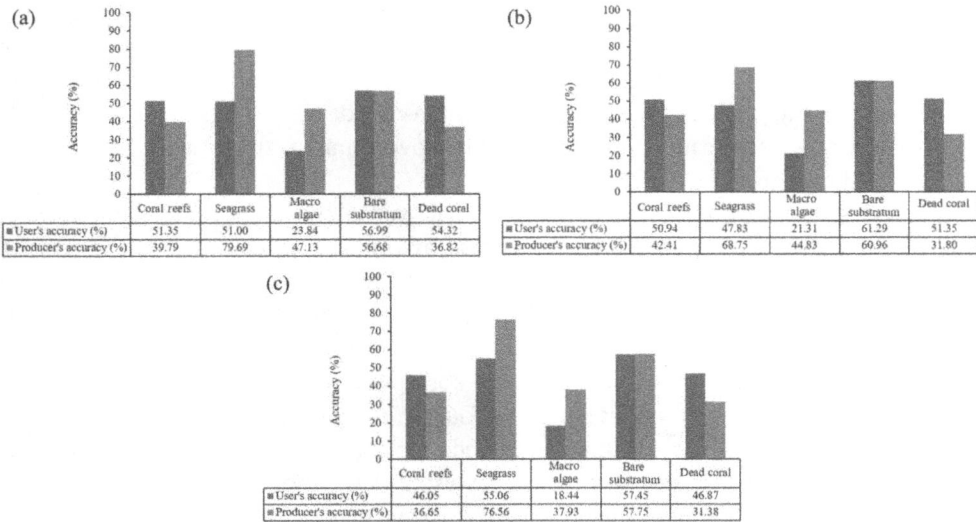

Figure 3. Comparison of the user's accuracy and producer's accuracy per input bands for benthic habitat mapping using per-pixel classification algorithms (a) BOA reflectance bands using CTA (Overall accuracy 47.13%, Kappa 0.33), (b) PC bands using CTA (Overall accuracy 46.09%, Kappa 0.32), and (c) MNF bands using CTA (Overall accuracy 43.61%, 0.29).

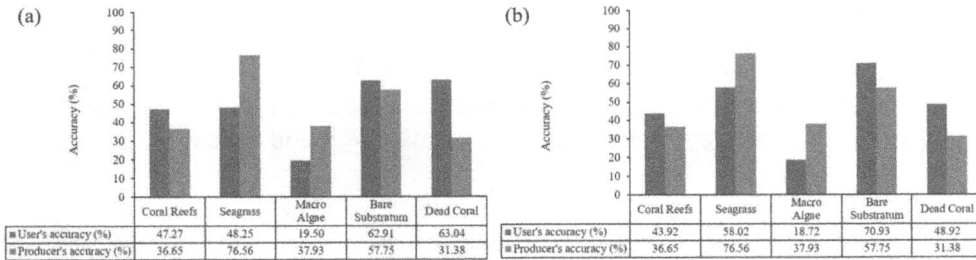

(a)

	Coral Reefs	Seagrass	Macro Algae	Bare Substratum	Dead Coral
▪ User's accuracy (%)	47.27	48.25	19.50	62.91	63.04
▪ Producer's accuracy (%)	36.65	76.56	37.93	57.75	31.38

(b)

	Coral Reefs	Seagrass	Macro Algae	Bare Substratum	Dead Coral
▪ User's accuracy (%)	43.92	58.02	18.72	70.93	48.92
▪ Producer's accuracy (%)	36.65	76.56	37.93	57.75	31.38

Figure 4. Comparison of user's accuracy and producer's accuracy of benthic habitat map obtained from OBIA using (a) BOA reflectance bands (Overall accuracy 46.84%, Kappa 0.33) and (b) MNF bands (Overall accuracy 46.22%, Kappa 0.32). The pattern of accuracy for each benthic class was similar for both OBIA results, where macroalgae class was the most difficult class to map, while seagrass and bare substratum were the most accurate classes.

OBIA of BOA reflectance bands (Figure 4) with the same value of seagrass PA (76.56%); hence, the OBIA result from MNF bands produced less-overestimated seagrass spatial distribution compared to OBIA of BOA reflectance bands. The UA and PA of seagrass class from per-pixel classification results were also less than OBIA of MNF result. Therefore, the seagrass mask to perform seagrass species mapping was adapted from OBIA result from MNF bands.

The application of OBIA using the integration of image segmentation and CTA did not produce benthic habitat map with higher OA. This can be partially addressed to the use of point samples to assess the accuracy of classification results. However, this issue could not be avoided as we did not have area-based benthic habitats *in situ* data. Hence, the spatial dimension of OBIA result, which is the advantage of OBIA, could not be fully assessed by our accuracy assessment process. Actually, extrapolating the point samples into area samples based on the segment used in the OBIA process can also be used, but it will result in the uncertainty of the *in situ* reference data. The *in situ* reference data used to test the classification accuracy should have minimal uncertainty to be confidently called a reference.

Coral reefs class was highly misclassified as macroalgae and dead coral, and vice versa (Table 5). Thus, the accuracy of these three classes was low as in per-pixel classification results. This can be explained by the spectra similarities and the confusion in the class descriptor between macroalgae and dead coral. Coral reefs, macroalgae (*in situ* data was dominated by brown macroalgae species), and dead coral have almost similar brownish colour. Thus, the pigment composition may be similar, hence the absorption feature. Furthermore, coral reefs and macroalgae may contain various types of colour pigments (Hochberg and Atkinson 2000). Dead coral overgrown by macroalgae also exhibit a similar spectral response to macroalgae and coral reefs. Labelling the dead coral class was also tricky, as most dead coral samples were associated with macroalgae. Although the dominant class in particular sample was dead coral, the associated-macro algae are the major contributor to the total reflectance of the corresponding pixel. Hence, the misclassification between macroalgae and dead coral might not be entirely classification error since pixels classified as dead coral might include macroalgae, such as turf algae and coralline algae, and vice versa.

Seagrass class was mostly misclassified to the bare substratum class, mainly because seagrass is highly associated with carbonate sand, which is their dominant substrate. For lower density seagrass, carbonate sand reflectance might have altered the resultant seagrass pixel reflectance significantly. This was also the reason for the misclassification of bare substratum class to coral reefs and dead coral class. Pixels of coral reefs or dead coral having lower coverage compared to the sand background were having a high risk

Table 5. Confusion matrix of the most accurate benthic habitat map from per-pixel classification and Object-based Image Analysis (OBIA) on PlanetScope image acquired on 17 May 2017. BOA: Bottom-of-Atmosphere, CTA: Classification Tree Analysis, UA: User Accuracy, PA: Producer Accuracy, OA: Overall Accuracy.

			Reference				
Per-pixel classification (CTA on BOA bands)							
Classified pixels	Coral reefs	Seagrass	Macro Algae	Bare Substratum	Dead Coral	Total	UA (%)
Coral reefs	76	0	18	17	37	148	51.35
Seagrass	9	51	5	20	15	100	51.00
Macroalgae	32	0	41	26	73	172	23.84
Bare Substratum	31	10	13	106	26	186	56.99
Dead Coral	43	3	10	18	88	162	54.32
Total	191	64	87	187	239		
PA (%)	39.79	79.69	47.13	56.68	36.82	OA 47.13% (0.33)	
OBIA (CTA on BOA bands)							
Coral reefs	78	0	23	10	54	165	47.27
Seagrass	17	55	0	16	26	114	48.25
Macroalgae	60	0	39	45	56	200	19.50
Bare Substratum	19	8	13	95	16	151	62.91
Dead Coral	17	1	12	21	87	138	63.04
Total	191	64	87	187	239		
PA (%)	36.65	76.56	37.93	57.75	31.38	OA 46.84% (0.33)	

of being classified as bare substratum since it has higher reflectance and overwhelmed the reflectance of the minor coral reefs or dead coral. Seagrass class was also misclassified as dead coral, especially with dead coral present near the shoreline.

It was clear that the extent of seagrass class was overestimated (Figure 5). Most misclassification occurred in back-reef and reef-crest area. Pixels in these two areas should be classified as coral reefs. Dead coral class was also stretching massively along reef-crest, which was unlikely as coral reefs in reef-crest across the study area were dominated by healthy coral reefs. Dead coral was correctly classified in areas near the shoreline associated with seagrass and macroalgae. Macroalgae were correctly classified in the shallow lagoon of the west part of Kemujan Island and in some areas near the shoreline and adjacent to seagrass. However, the occurrence of macroalgae class in the reef-crest and fore-reef indicating misclassification. To sum up, coral reefs, macroalgae, and dead coral class were highly misclassified. This occurred mainly in the back-reef and reef-crest area where dead coral and macroalgae should not be the dominant benthic cover.

4.1.2. Comparison of May and August 2017 image

The accuracy of PlanetScope image acquired on 15 August 2017 was relatively similar to the accuracy of PlanetScope image acquired in May 2017 despite the samples differences. The highest OA was 50% for August image obtained from CTA of PC bands. The UA and PA for each benthic class from August result had a slightly different pattern, where the UA of the dead coral class was higher than the PA as opposed to the May result. Seagrass and bare substratum class were classified at higher accuracy in May result, while coral reefs and macroalgae class had a higher accuracy in August result. The dead coral class was classified with relatively similar accuracy.

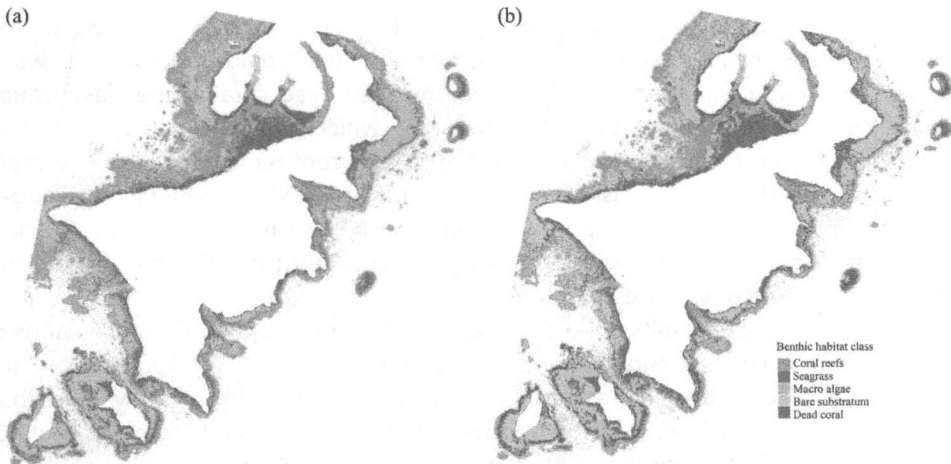

Figure 5. Map of benthic habitats obtained from (a) per-pixel CTA using MNF bands with 43.61% overall accuracy (OA). Seagrass class user's accuracy (UA) and producer's accuracy (PA) were 55.06% and 76.56%, respectively, (b) OBIA using CTA using MNF bands (46.22% OA). Seagrass class UA and PA were 58.02% and 76.56%, respectively. These two results produced the highest seagrass UA and PA compared to other classification results despite not having the highest accuracy.

Table 6. Confusion matrix of benthic habitat classification of August result. UA: User Accuracy, PA: Producer Accuracy, OA: Overall Accuracy.

	Reference						
Per-pixel classification (CTA on PC bands)							
Classified pixels	Coral reefs	Seagrass	Macro Algae	Bare Substratum	Dead Coral	Total	UA (%)
Coral reefs	103	2	4	10	32	151	68.21
Seagrass	13	44	19	25	3	104	42.31
Macroalgae	25	8	46	43	7	129	35.66
Bare Substratum	15	9	18	79	7	128	61.72
Dead Coral	54	2	6	18	48	128	37.50
Total	210	65	93	175	97		
PA (%)	49.05	67.69	49.46	45.14	49.48	**OA 50% (0.37)**	
OBIA (CTA on PC bands)							
Coral reefs	97	11	10	8	27	153	63.40
Seagrass	5	35	13	17	3	73	47.95
Macroalgae	17	10	31	29	6	93	33.33
Bare Substratum	29	9	28	96	8	170	56.57
Dead Coral	62	0	11	25	53	151	35.10
Total	210	65	93	175	97		
PA (%)	46.19	53.85	33.33	54.86	54.64	**OA 48.75% (0.34)**	

The pattern of misclassification for May and August image was also relatively similar (Table 6). Coral reefs and dead coral class were highly misclassified. The rate of misclassification of macroalgae class was lower with dead coral but higher with the bare substratum. The misclassification between coral reefs and macroalgae was still present. Seagrass class was also still misclassified highly as bare substratum. However, the rate of misclassification between seagrass and macroalgae was increasing. These results were a good indicator of the general performance of PlanetScope image for benthic habitat mapping at this level of complexity. Direct comparison with previous works could not be easily made as others did not explicitly include a dead coral class in the scheme, hence producing higher accuracy, i.e., Zhang et al. (2013), Eugenio, Marcello, and Martin (2015), Wicaksono (2016). Awak et al. (2016) reported 47% OA (excluding optically deep water class) for mapping coral reefs ecosystem that involved dead coral in the classification scheme. We produced maps with 40–50% accuracy, which was comparable to previous works. This relatively low OA was mainly due to the confusion between coral reefs, macroalgae, and dead coral class in addition to the low SNR of PlanetScope images. Kutser, Miller, and Jupp (2006) reported the effectiveness of mapping dead coral class using hyperspectral image using the spectral library as input endmember but did not report the actual accuracy assessment.

In addition, from the results of both dates, it appeared that BOA reflectance bands of PlanetScope image were accurate enough to map benthic habitat. For August result, the accuracy of BOA reflectance bands was 48.9% (0.35), which was only slightly lower than the best OA. The application of image transformation may improve the accuracy but not significantly, and the accuracy improvement is not always obtained.

4.2. Seagrass species mapping

4.2.1. Seagrass species mapping of 17 May 2017 image

Seagrass species map with the highest OA was obtained from CTA, where any inputs produced OA > 70% (Table 7). Seagrass species fraction from LSU produced the highest accuracy with 74.31% OA, which is 1–2% higher than PC bands and BOA reflectance bands OA. SVM and ML followed with 50–66% OA. SVM produced higher OA than ML, however, not all seagrass species were able to be classified by SVM, hence lower Kappa value. ML was more consistent in term of mapping different seagrass species compared to SVM. SAM and SID produced very low accuracy with 3.77% and 13.91% OA, respectively. The low accuracy can be addressed to the similarity of spectra shape between seagrass species and the impact of water column energy attenuation, as encountered in our previous work using the WorldView-2 image (Wicaksono et al. 2017).

Different inputs produced different patterns of seagrass species UA and PA (Figure 6). The extent of *CrHu* class was slightly overestimated in BOA reflectance bands, PC bands, and seagrass species fraction result. MNF bands result has slightly underestimated the extent of *CrHu* but the chance of pixel classified as *CrHu* in the image is *CrHu* in the field was the best with 64.29%. The extent of *Ea* class was underestimated for all input except for MNF bands. *EaThCr* class extent was over-estimated for all input but seagrass species fraction. *Th* class was failed to be properly classified using MNF bands, and the pattern for other inputs was also varied. The extent of *Th* class in BOA reflectance bands result was greatly underestimated while in PC Bands and seagrass species fraction was highly overestimated. Only *ThCr* class has a similar pattern for all inputs.

The result of seagrass species mapping using OBIA was lower (69.06%) due to the same issue encountered during benthic habitat mapping. This was obtained from OBIA with CTA using PC bands. The accuracy for individual species class was similar where *ThCr* class was mapped with the highest accuracy (UA 79.60%, PA 87.67%). *Th* class was highly overestimated (UA 35.48%, PA 91.67%). *EaThCr* class was slightly overestimated (UA 44.00%, PA 50.00%). *Ea* (UA 45.00%, PA 30.00%) and *CrHu* class (UA 55.56%, 28.17%) were underestimated.

The misclassification rate between seagrass species class can be seen in Table 8. Most misclassification occurred between *CrHu* and *ThCr* class, which might be caused by the presence of *Cr* species in both classes. *EaThCr* class was mostly misclassified as *Ea* or *ThCr*

Table 7. Summary of accuracy assessment of May PlanetScope image for seagrass species mapping. The use of PlanetScope BOA reflectance bands were adequate to map seagrass species when proper classification algorithm was used. Values in bold are the highest accuracy for each input.

	Overall Accuracy (OA) of Seagrass Species Mapping							
	BOA reflectance bands		PC bands		MNF bands		Seagrass fraction	
Algorithm	OA (%)	Kappa	OA (%)	Kappa	OA (%)	Kappa	OA (%)	Kappa
ML	58.57	0.31	62.38	0.35	62.38	0.35	0.00	0.00
SVM	63.91	0.13	64.85	0.17	64.85	0.17	65.09	0.18
CTA	**72.65**	**0.53**	**73.48**	**0.56**	**70.99**	**0.51**	**74.31**	**0.56**
SAM	3.77	0.01						
SID	13.91	0.03						

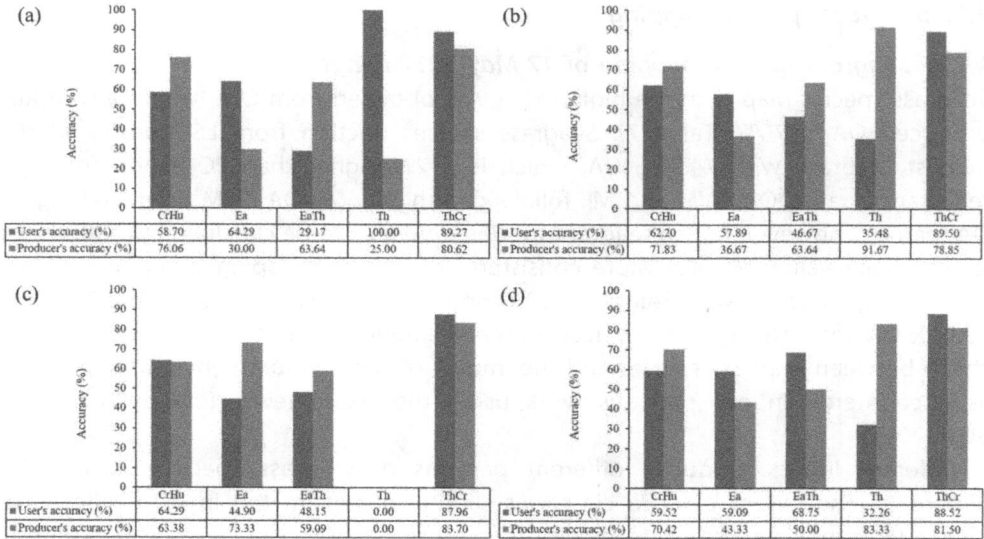

Figure 6. Comparison of user's accuracy and producer's accuracy of seagrass species mapping of May image (a) BOA reflectance bands using CTA (Overall accuracy 72.65%, Kappa 0.53), (b) PC bands using CTA (Overall accuracy 73.48%, Kappa 0.56), (c) MNF bands using CTA (Overall accuracy 70.99%, Kappa 0.51), and (d) Seagrass fraction from LSU using CTA (Overall accuracy 74.31%, Kappa 0.56).

and vice versa, which was understandable as *EaThCr* class is a mixed class consist of *Ea* and *ThCr* classes. *Th* class, which was classified inconsistently between inputs, was mainly *Ea* in the field.

Seagrass species maps from May image are shown in Figure 7. The spatial distribution of seagrass species in the map adequately represents the actual condition in the study area, except in areas where seagrass class was misclassified as other benthic classes as discussed previously. *Ea* and *EaThCr* class dominates water in the western part of the study area. Meanwhile, *Th, ThCr,* and *CrHu* class are more common in the eastern part. However, noticeable misclassification occurred in the reef-flat in the western part of mangrove forest (see blue boxes in Figure 7). Many pixels in this area were classified as either *Ea* or *EaThCr* class. In fact, this area is dominated by low coverage of *Th* or *Cr*, with the very limited spatial distribution of *Ea*. It was suspected that many pixels were misclassified as *EaThCr* class because of the presence of *Th* and *Cr*.

4.2.2. *Comparison with 15 August 2017 image*

The highest OA for August image was 74.03% (0.547) and 65.74% (0.324) for CTA per-pixel classification (seagrass species fraction) and OBIA (PC bands) respectively. Other classification algorithms produced significantly lower accuracies: SVM 64.39%, ML 43.11%, SAM 4.76%, and SID 10.27%. This was highly comparable to the May results. The misclassification pattern was also similar (see Table 8), where the highest misclassification occurred between *CrHu* and *ThCr* class. *EaThCr* class was mainly misclassified as either *Ea* or *ThCr*. *Th* class was still inconsistently classified with no pattern of accuracy or misclassification. This indicated that the spatial distribution of seagrass species in the study area in May and August was similar since the use of training area obtained in

Table 8. Confusion matrix of seagrass species classification of May and August image. Refer to Table 2 for seagrass species code explanation. UA: User's Accuracy, PA: Producer's Accuracy, OA: Overall Accuracy.

			Reference				
May: Per-pixel classification (CTA on seagrass fraction from LSU)							
Classified pixels	CrHu	Ea	EaThCr	Th	ThCr	Total	UA (%)
CrHu	50	1	1	1	31	84	59.52
Ea	3	13	2	0	4	22	59.09
EaThCr	0	3	11	0	2	16	68.75
Th	1	12	3	10	5	31	32.26
ThCr	17	1	5	1	185	209	88.52
Total	71	30	22	12	227		
PA (%)	70.42	43.33	50.00	83.33	81.50	OA 74.31% (0.56)	
May: OBIA (CTA on PC bands)							
CrHu	20	1	1	0	14	36	55.56
Ea	2	9	5	0	4	20	45.00
EaThCr	2	6	11	1	5	25	44.00
Th	0	12	3	11	5	31	35.48
ThCr	47	2	2	0	199	250	79.60
Total	71	30	22	12	227		
PA (%)	28.17	30.00	50.00	91.67	87.67	OA 69.06% (0.42)	
August: Per-pixel classification (CTA on seagrass fraction from LSU)							
CrHu	50	1	1	0	36	88	56.82
Ea	0	24	5	11	6	46	52.17
EaThCr	0	3	13	0	5	21	61.90
Th	0	0	0	0	0	0	0.00
ThCr	21	2	3	1	180	207	86.96
Total	71	30	22	12	227		
PA (%)	70.42	80.00	59.09	0.00	79.30	OA 74.03% (0.55)	
August: OBIA (CTA on PC bands)							
CrHu	8	6	5	0	12	31	25.81
Ea	1	16	6	7	5	35	45.71
EaThCr	3	4	8	1	8	24	33.33
Th	0	0	0	4	0	4	100
ThCr	59	4	3	0	202	268	75.37
Total	71	30	22	12	227		
PA (%)	11.27	53.33	36.36	33.33	88.99	OA 65.74% (0.32)	

August can be successfully used in May image without any loss of accuracy. Also, this also justified the consistency of quality and performance of PlanetScope images for seagrass species mapping.

5. Discussion

PlanetScope image is a new image with many advantages, including high temporal resolution obtained at high spatial resolution. The spatial resolution of PlanetScope image is comparable to WorldView-2 (1.84 m), IKONOS (4 m), Quickbird (2.4 m), and Rapideye (5 m), as well as its spectral resolution. Hence, with this specification, PlanetScope image appears to be promising for extracting natural resources information, including benthic habitat and seagrass biodiversity. Among these satellite images, Quickbird, WorldView-2, and IKONOS were the most successful in case of benthic habitat and seagrass mapping (Andréfouët et al. 2003; Phinn, Roelfsema, and Mumby 2012; Lyons, Roelfsema, and Phinn 2013; Roelfsema et al. 2014). For instance, recent

(a)   (b)

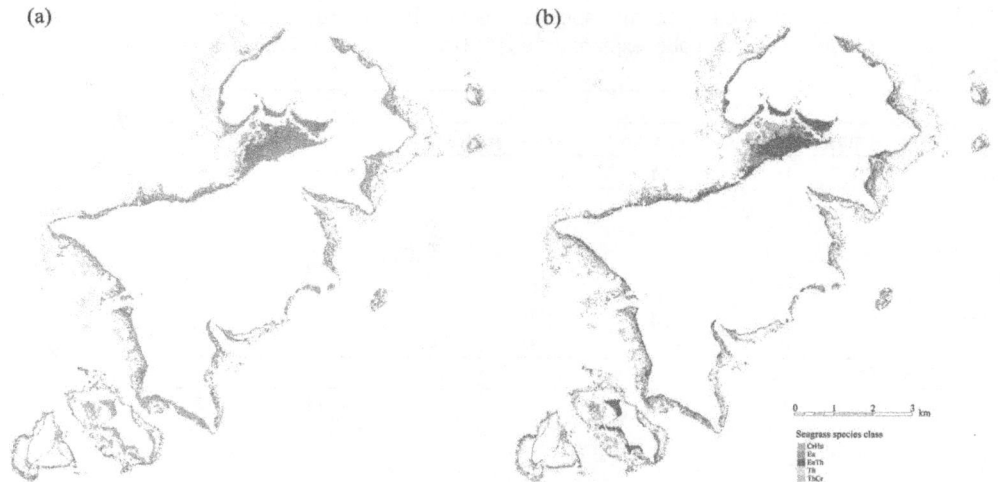

Figure 7. Map of seagrass species from May PlanetScope image (a) CTA per-pixel classification using seagrass fraction from LSU with 74.31% overall accuracy and Kappa 0.56 and (b) OBIA using PC bands with 69.06% overall accuracy and Kappa 0.42. Blue boxes indicate the area where noticeable misclassification occurred.

publications of benthic habitats mapping revealed that multispectral images such as WorldView-2 produced accuracy more than 60% (Eugenio, Marcello, and Martin 2015; Wicaksono 2016), however, none of these works included dead coral class. SPOT 6, SPOT 7 and Rapideye were rarely used to map benthic habitat and seagrass, but Awak et al. (2016) delivered 47% accuracy using Rapideye for benthic habitat mapping.

We are among the first to test the performance of PlanetScope image for underwater objects mapping, specifically benthic habitat including seagrass species. The accuracy of PlanetScope image was relatively good for mapping benthic habitat at five classes complexity, although the accuracy is still lower compared to other high spatial resolution image such as Quickbird (Mishra et al. 2006; Phinn, Roelfsema, and Mumby 2012), IKONOS (Capolsini et al. 2003), WorldView-2 (Eugenio, Marcello, and Martin 2015; Wicaksono 2016), and Rapideye (Awak et al. 2016). Based on the Kappa statistic (Landis and Koch 1977), the strength of agreement for benthic habitat mapping is Fair (0.21–0.40). However, Phinn, Roelfsema, and Mumby (2012) managed to obtain much higher accuracy for mapping complex benthic habitat using a hierarchical approach up to reef benthic community level using Quickbird. In Phinn, Roelfsema, and Mumby (2012), the live and dead coral class were combined. Mishra et al. (2006), Eugenio, Marcello, and Martin (2015) also did not include dead coral class in the classification scheme. Nevertheless, Capolsini et al. (2003), Awak et a. (2015), and Wicaksono (2016) explicitly used dead coral in the classification scheme and still produced higher accuracy. We argued that this was also caused by the difference in image quality, especially the low SNR of PlanetScope image, the classification approached used, and the use of dead coral class.

For seagrass species mapping, the accuracy of PlanetScope image was also relatively good with more than 70% OA and Moderate agreement (0.41–0.60) for five classes seagrass species mapping. Seagrass species mapping with seven species classes using per-pixel classification algorithm such as Minimum Distance (MD) delivered low accuracy

(<30%) even when the high-spatial resolution hyperspectral image was used (Phinn et al. 2008). A direct comparison could not be made since the only similar species was *Ho*, and in our study area, *Ho* did not create a bed. It was only a minor species found between the more dominant and bigger species or found in a very sparse density in an area dominated by carbonate sand substrate. *Ho* was commonly found in a mixed bed together with *Hu*, *Cr*, or *Th* at a very sparse density in the north coast of Ujung Gelam. However, *Padina sp.* and *Sargassum sp.* were also present on the same bed, and since they are bigger and had higher coverage than *Hu* and *Ho*, and the reflectance of carbonate sand also contribute strongly, we could not incorporate *HuHo* bed as a training area. The results from Phinn et al. (2008) and Roelfsema et al. (2014) showed that seagrass class involving *Ho* was the least accurate. Other species such as *Cs* and *Si* were limitedly found in our study area. Our results were also comparable to the accuracy of seagrass species mapping using the hyperspectral image (Pu et al. 2012; Pu, Bell, and English 2015), which obtained 70–77% accuracy for mapping three seagrass species. Lyons, Phinn, and Roelfsema (2011) managed to produce five classes of seagrass species map with 62% and 80% accuracy using Quickbird with ML, which was also similar to our ML accuracy.

Roelfsema et al. (2014) obtained much higher accuracy using OBIA to map five seagrass species multi-temporally using Quickbird and WorldView-2. Most of the time, the accuracy of individual seagrass species was above 60% (both UA and PA). Only the mean PA of *Ho* is below 50%. The individual species class accuracy and OA were consistent across time and seasons, with a mean OA of 77% and the accuracy of some species exceeded 80%. In our research, we obtained better accuracy using per-pixel machine-learning algorithm CTA than OBIA, mainly due to the point samples used to assess the accuracy. We also obtained more than 70% accuracy but less consistent in term of the UA and PA of individual seagrass species class, i.e., *Th* class. However, it is important to note that this statistical accuracy might not be directly comparable despite the species similarity. The complexity of seagrass beds, species mixing, and species composition was unique for each study area, and this sets the difference in the level of difficulty in mapping seagrass species.

Another issue that occurred during seagrass species mapping was the process of accuracy assessment, especially for mixed species class such as *CrHu*, *ThCr*, and *EaThCr*. Mixed class means that several species present in the field. For instance, if a pixel classified as *ThCr* was only *Th* or *Cr* in the field, the confusion matrix rule considered this as a misclassification, hence calculated as an error. However, in fact, it is debatable to consider this as a real misclassification, as half of the classified class was correct. This issue will escalate when using lower spatial resolution image where finding pixel with real homogenous seagrass species is difficult. Therefore, it is crucial to develop accuracy assessment method that fairly assesses this mixed class accuracy. The seagrass species classification scheme must also be supported by the correct class descriptor, i.e., whether particular minor species may present. Moreover, whether *ThCr* class means both *Th* and *Cr* are present or meaning that *Th* and or *Cr* are present must be carefully defined.

In this study, we applied different classification algorithms to PlanetScope images, from the commonly used ML to the machine-learning image mining approach and their combination with image segmentation. The results suggested that for benthic habitat and seagrass species mapping, machine-learning CTA delivered the best accuracy, not only the OA but also the individual class accuracy. The accuracy difference of CTA results between inputs was small compared other classification algorithms, and each class was consistently mapped with relatively similar accuracy despite the inputs. Therefore, we recommend the use of CTA for benthic habitat and seagrass species mapping. Other algorithms such as SVM and ML produced lower and less consistent accuracy. The accuracy of SAM and SID for seagrass mapping was highly affected by the similarity of seagrass species spectra and the effect of the water column, which provide gaps with the seagrass species endmember obtained from field spectroscopy. However, we argue that although water column effect is compensated for, the increase in SAM or SID accuracy will not be significant due to the limited number of the spectral band of PlanetScope image to make these two algorithms work effectively.

The accuracy of PlanetScope images for benthic habitat mapping between dates was relatively consistent, which is a good indicator of consistent radiometric quality. The use of BOA reflectance image without additional image transformation and correction delivered promising accuracy. Although more study is required to set the standard performance of PlanetScope image, our study provided some important accuracy base-line. Thus, with the benefit of high temporal revisit, PlanetScope image is significantly better for mapping and monitoring benthic habitat changes and dynamics.

However, despite the positive, PlanetScope image is not without issue. The noise level is high, and the radiometric quality and inconsistency are low on homogeneous clusters of pixels, especially in homogenous optically deep water pixels. This is an indication of low SNR of PlanetScope image. This issue was encountered during the process of sunglint correction. To perform sunglint correction, a strong correlation between visible and NIR bands must exist for optically deep water pixel affected by sunglint. Previous researches by Hedley, Harborne, and Mumby (2005) and Kay, Hedley, and Lavender (2009) have already documented the positive correlation between visible and NIR bands with a very strong coefficient of determination (>0.8). Compared with PlanetScope image, it was not possible to obtain a good correlation between visible and NIR band. The correlation was mostly below 0.4 in all combination of visible and NIR band. We also tried to select different samples across the optically deep water pixels in April and August image, but we still failed to obtain strong correlation. In addition to the low correlation, the nature of the relationship between visible and NIR band was not consistent. According to Mobley (1994), the index of refraction between visible and NIR band is almost similar and positive correlation is expected. In our case, the correlation was sometimes positive and negative, which indicated the radiometric inconsistencies of PlanetScope image over optically deep water pixels. This lead to the consequence of not being able to correct sunglint from PlanetScope image.

In our case where sunglint was not acutely occurring, this issue might not raise a significant concern. However, if sunglint is strong and significantly limit the ability to obtain underwater information, the failure to perform sunglint correction may reduce the effectiveness of mapping using PlanetScope image. This is something that must be addressed by PlanetScope engineer. In comparison, based on our experience, Rapideye also had a similar issue where it was very difficult to perform sunglint correction due to the low correlation between visible and NIR band.

Finally, it is positively appreciated that we can have remote-sensing image coincide with the date of field survey, which is so far, the issue seriously encountered in the previous works of remote-sensing-based mapping. The use of the high-spatial-resolution image (<5 m) acquired close to the date of field survey is critical to obtain better seagrass species mapping accuracy (Phinn et al. 2008). Especially in the tropics, this will be a great benefit since cloud-cover is high and find a cloud-free image is difficult. Daily image acquisition will provide us with an ideal image alternative if the condition during field survey is not ideal for remote-sensing image acquisition. Planet is also currently releasing the PlanetScope SR product, which makes the processing becomes faster. This SR product is important since the PlanetScope image is provided on a daily basis and radiometric calibration is required to make use of this multi-temporal images effectively. Having images on a daily basis means that the variation due to atmospheric conditions presents and making them atmospheri-cally corrected is necessary. Processing such huge daily data require too much efforts if we have to perform the correction manually for each scene. The availability of SR product become another plus for Planet constellation data as compared to other high and medium resolution satellite images. Landsat series has already released their Analysis-Ready Data (ARD) level that allows us to directly use the image for analysis without having to perform atmospheric and geometric correction. Time-series comparison of earth-surface information at the different level of precisions will become easier than ever if we can combine both Landsat and Planet constellation data, with the expectation that Sentinel-2 will also make its ARD level available. Monitoring changes in benthic habitats can be made more precise with PlanetScope image availability, especially for the immediate response of any occurrence of an adverse event caused by natural phenomena such as storm, disaster, or anthropogenic such as shipping, tourism, fishing activities.

6. Conclusions

We presented the first assessment of PlanetScope multi-temporal images for benthic habitat and seagrass species mapping. Several significant findings regarding the performance of PlanetScope image is as follows. PlanetScope image is a consistent performer. The accuracy between images is relatively consistent, and the relatively good accuracy can be obtained without applying additional transformation or corrections given the appropriate image classification method. Furthermore, PlanetScope satellites obtain images at almost daily basis, hence allows us to obtain remote-sensing image very close to the date of field survey. As a result, the temporal bias is minimal, and highly beneficial for mapping dynamic objects and mapping in tropical regions. Although the accuracy is not directly comparable to previous works due to the difference in the classification scheme used and complexity of the benthic habitat, the accuracy of seagrass species mapping (74.31%) of PlanetScope image are com-parable to those from Quickbird, IKONOS, and WorldView-2. Indeed, the accuracy of benthic habitat mapping (50.00%) is lower than other similar images such as Quickbird, IKONOS, and WorldView-2 and Rapideye. It also has SR product available to users, which is strongly required to obtain a radiometrically comparable image across time and space. Nevertheless, PlanetScope image has a serious issue regarding the reflectance of optically deep water pixels, which has low SNR and limits the application of sunglint correction. Mapping benthic habitat and seagrass species in water with high sunglint will be very challenging and difficult. This is a serious issue that must be addressed by Planet engineers for future PlanetScope satellites development.

Acknowledgments

This is part of our feedback in our involvement in the Planet Education and Research program (Planet Team (2017). Planet Application Program Interface: In Space for Life on Earth. San Francisco, CA. https://api.planet.com). We would like to thank Planet Team for the opportunity to use and assess the performance of PlanetScope images.

Disclosure statement

No potential conflict of interest was reported by the authors.

Funding

The measurement of seagrass species reflectance spectra was funded by "Direktorat Riset dan Pengabdian Masyarakat – Direktorat Jenderal Penguatan Riset dan Pengembangan – Kementerian Riset, Teknologi, dan Pendidikan Tinggi Republik Indonesia" via Fundamental Research Scheme Grant Number 2232UN1-P.III/DIT-LIT/LT/2017.

References

Andréfouët, S., P. Kramer, D. Torres-Pulliza, K. E. Joyce, E. J. Hochberg, R. Garza-Perez, F. E. Muller-Karger, et al. 2003. "Multi-Site Evaluation of IKONOS Data for Classification of Tropical Coral Reef Environments." *Remote Sensing of Environment* 88 (1–2): 128–143. doi:10.1016/j.rse.2003.04.005.

Awak, D. S., J. L. Gaol, B. Subhan, H. H. Madduppa, and D. Arafat. 2016. "Coral Reef Ecosystem Monitoring Using Remote Sensing Data: Case Study in Owi Island, Biak, Papua." *Procedia Environmental Sciences* 33: 600–606. Elsevier. doi:10.1016/j.proenv.2016.03.113.

Ayoobi, I., and M. H. Tangestani. 2017. "The Effect of Minimum Noise Fraction Data Input on Success of Artificial Neural Network in Lithological Mapping of A Magmatic Terrain with ASTER Data: A Case Study from SE Iran." *Remote Sensing Applications: Society and Environment* 7: 21–26. doi:10.1016/j.rsase.2017.06.001.

Bell, S., M. Fonseca, and N. B. Stafford. 2006. "Segrass Ecology: New Contributions from a Landscape Perspective." In *Seagrasses: Biology, Ecology and Conservation*, edited by A. W. Larkum, R. Orth, and C. M. Duarte. Dordrecht, The Netherlands: Springer.

Benz, U. C., P. Hofmann, G. Willhauck, M. Lingenfelder, and M. Heynen. 2004. "Multi-Resolution, Object-Oriented Fuzzy Analysis of Remote Sensing Data for GIS-Ready Information." *ISPRS Journal of Photogrammetry & Remote Sensing* 58: 239–258. doi:10.1016/j.isprsjprs.2003.10.002.

Capolsini, P., S. Andrefouet, C. Rion, and C. Payri. 2003. "A Comparison of Landsat ETM+, SPOT HRV, IKONOS, ASTER and Airborne MASTER Data for Coral Reef Habitat Mapping in South Pacific Island." *Canadian Journal of Remote Sensing* 29 (2): 187–200. doi:10.5589/m02-088.

Casal, G., T. Kutser, J. A. Dominguez-Gomez, N. Sanchez-Carnero, and J. Freire. 2011. "Mapping Benthic Macroalgal Communities in the Coastal Zone Using CHRIS-PROBA Mode 2 Images." *Estuarine, Coastal and Shelf Science* 94: 281–290. doi:10.1016/j.ecss.2011.07.008.

Collison, A., and N. Wilson. 2017. *Planet Surface Reflectance Product*. San Francisco, CA: Planet Labs.

Congalton, R. G., and K. Green. 2008. *Assessing the Accuracy of Remotely Sensed Data: Principles and Practices. Mapping Science*. Boca Rotan, FL: CRC Press.

Dekker, A., V. Brando, J. Anstee, S. K. Fyfe, T. Malthus, and E. Karpouzli. 2006. "Remote Sensing of Seagrass Systems: Use of Spaceborne and Airborne Systems." In *Seagrasses: Biology, Ecology and Conservation*, edited by A. Larkum, R. Orth, C. Duarte, and C. Duarte, 347–359. Dordrecht: Springer.

Duffy, J. E. 2006. "Biodiversity and the Functioning of Seagrass Ecosystems." *Marine Ecology Progress Series* 311: 233–250. doi:10.3354/meps311233.

Eugenio, F., J. Marcello, and J. Martin. 2015. "High-Resolution Maps of Bathymetry and Benthic Habitats in Shallow-Water Environments Using Multispectral Remote Sensing Imagery." *IEEE Transactions on Geoscience and Remote Sensing* 53 (7): 3539–3549. doi:10.1109/TGRS.2014.2377300.

Goodman, J. A., S. J. Purkis, and S. R. Phinn. 2013. *Coral Reef Remote Sensing A Guide for Mapping, Monitoring and Management.* Edited by S. R. Phinn. Dordrecht: Springer.

Green, E. P., P. J. Mumby, A. J. Edwards, and C. D. Clark. 2000. *Remote Sensing Handbook for Tropical Coastal Management. Coastal Management Sourcebooks 3.* (A. J. Edwards, Ed.). Paris: UNES.

Hedley, J. D., A. R. Harborne, and P. J. Mumby. 2005. "Simple and Robust Removal of Sunglint for Mapping Shallow-Water Benthos." *International Journal of Remote Sensing* 26 (10): 2107–2112. doi:10.1080/01431160500034086.

Hedley, J. D., B. J. Russel, K. Randolph, M. A. Castro, R. M. Elizondo, S. Enriquez, and H. M. Dierssen. 2017. "Remote Sensing of Seagrass Leaf Area Index and Species: The Capability of a Model Inversion Method Assessed by Sensitivity Analysis and Hyperspectral Data of Florida Bay." *Frontiers in Marine Science* 4. doi:10.3389/fmars.2017.00362.

Hedley, J. D., C. M. Roelfsema, S. R. Phinn, and P. J. Mumby. 2012. "Environmental and Sensor Limitations in Optical Remote Sensing of Coral Reefs: Implications for Monitoring and Sensor Design." *Remote Sensing* 4: 271–302. doi:10.3390/rs4010271.

Hochberg, E. J., and M. J. Atkinson. 2000. "Spectral Discrimination of Coral Reef Benthic Communities." *Coral Reefs* 19: 164–171. doi:10.1007/s003380000087.

Hossain, M. S., J. S. Bujang, M. H. Zakaria, and M. Hashim. 2015. "The Application of Remote Sensing to Seagrass Ecosystems: An Overview and Future Research Prospects." *International Journal of Remote Sensing* 36 (1): 61–113. doi:10.1080/01431161.2014.990649.

Huang, C., L. S. Davis, and J. R. Townshend. 2002. "An Assessment of Support Vector Machines for Land Cover Classification." *International Journal of Remote Sensing* 23: 725–749. doi:10.1080/01431160110040323.

Kay, S., J. D. Hedley, and S. Lavender. 2009. "Sun Glint Correction of High and Low Spatial Resolution Images of Aquatic Scenes: A Review of Methods for Visible and Near-Infrared Wavelengths." *Remote Sensing* 1: 697–730. doi:10.3390/rs1040697.

Kenworthy, W. J., S. Wyllie-Echeverria, R. G. Coles, G. Pergent, and C. Pergent-Martini. 2006. "Seagrass Conservation Biology: An Interdisciplinary Science for Protection of the Seagrass Biome." In *Seagrasses: Biology, Ecology and Conservation,* edited by W. D. Larkum, R. Orth, and C. M. Duarte, 595–623. Dordrecht, Netherlands: Springer.

Kohler, K. E., and S. M. Gill. 2006. "Coral Point Count with Excel Extensions (Cpce): A Visual Basic Program for the Determination of Coral and Substrate Coverage Using Random Point Count Methodology." *Computers & Geosciences* 32: 1259–1269. doi:10.1016/j.cageo.2005.11.009.

Kritzer, J. P., M. B. DeLucia, E. Greene, C. Shumway, M. F. Topolski, J. Thomas-Blate, L. A. Chiarella, K. B. Davy, and K. Smith. 2016. "The Importance of Benthic Habitats for Coastal Fisheries." *Bioscience* 66 (4): 274–284. doi: 10.1093/biosci/biw014.

Kutser, T., I. Miller, and D. L. Jupp. 2006. "Mapping Coral Reef Benthic Substrates Using Hyperspectral Space-Borne Images and Spectral Libraries." *Estuarine, Coastal and Shelf Science* 70: 449–460. doi:10.1016/j.ecss.2006.06.026.

Landis, J. R., and G. G. Koch. 1977. "The Measurement of Observer Agreement for Categorical Data." *Biometrics* 33 (1): 159–174.

Larkum, A. W., R. Orth, and C. M. Duarte. 2006. *Seagrasses: Biology, Ecology and Conservation.* Dordrecht, The Netherlands: Springer.

Lyons, M., S. Phinn, and C. Roelfsema. 2011. "Integrating Quickbird Multi-Spectral Satellite and Field Data: Mapping Bathymetry, Seagrass Cover, Seagrass Species and Change in Moreton Bay, Australia in 2004 and 2007." *Remote Sensing* 3: 42–64. doi:10.3390/rs3010042.

Lyons, M. B., C. M. Roelfsema, and S. R. Phinn. 2013. "Towards Understanding Temporal and Spatial Dynamics of Seagrass Landscapes Using Time-Series Remote Sensing." *Estuarine, Coastal and Shelf Science* 120: 42–53. doi:10.1016/j.ecss.2013.01.015.

McKenzie, L., M. A. Finkbeiner, and H. Kirkman. 2001. "Seagrass Mapping Methods." In *Global Seagrass Research Methods*, edited by F. T. Short and R. G. Coles. Amsterdam: Elsevier.

Mishra, D., S. Narumalani, D. Rundquist, and M. Lawson. 2006. "Benthic Habitat Mapping in Tropical Marine Environments Using QuickBird Multispectral Data." *Photogrammetric Engineering & Remote Sensing* 72 (9): 1037–1048. doi:10.14358/PERS.72.9.1037.

Mobley, C. D. 1994. *Light and Water: Radiative Transfer in Natural Waters*. Sand Diego: Academic Press.

Mumby, P. J., and E. P. Green. 2000. "Mapping Seagrass Beds." In *Remote Sensing Handbook for Tropical Coastal Management*, edited by A. J. Edwards, 175–182. Paris: UNES.

Nordlund, L. M., E. W. Koch, E. B. Barbier, and J. C. Creed. 2016. "Seagrass Ecosystem Services and Their Variability across Genera and Geographical Regions." *PLoS One* 11 (10): e0163091. doi:10.1371/journal.pone.0163091.

Phinn, S. R., C. M. Roelfsema, A. Dekker, V. Brando, and J. Anstee. 2008. "Mapping Seagrass Species, Cover and Biomass in Shallow Waters: An Assessment of Satellite Multi-spectral and Airborne Hyper-spectral Imaging Systems in Moreton Bay (Australia)." *Remote Sensing of Environment* 112: 3413–3425. doi: 10.1016/j.rse.2007.09.017.

Phinn, S. R., C. M. Roelfsema, and P. J. Mumby. 2012. "Multi-Scale, Object-Based Image Analysis for Mapping Geomorphic and Ecological Zones on Coral Reefs." *International Journal of Remote Sensing* 33 (12): 3768–3797. doi:10.1080/01431161.2011.633122.

Planet. 2017. *Planet Imagery Product Specification*. San Francisco, CA: Planet Labs.

PlanetTeam. 2017. *Planet Application Program Interface: In Space for Life on Earth*. San Francisco, CA. https://api.planet.com

Pu, R., S. Bell, C. Meyer, L. Baggett, and Y. Zhao. 2012. "Mapping and Assessing Seagrass along the Western Coast of Florida Using Landsat TM and EO-1 ALI/Hyperion Imagery." *Estuarine, Coastal and Shelf Science* 115: 234–245. doi:10.1016/j.ecss.2012.09.006.

Pu, R., S. Bell, and D. English. 2015. "Developing Hyperspectral Vegetation Indices for Identifying Seagrass Species and Cover Classes." *Journal of Coastal Research* 31: 595–615. doi:10.2112/JCOASTRES-D-12-00272.1.

Richards, J. A. 2013. *Remote Sensing Digital Image Analysis*. Berlin,Germany: Springer-Verlag.

Roelfsema, C. M., M. Lyons, E. M. Kovacs, P. Maxwell, M. I. Saunders, J. Samper-Villarreal, and S. R. Phinn. 2014. "Multi-Temporal Mapping of Seagrass Cover, Species and Biomass: A Semi-Automated Object Based Image Analysis Approach." *Remote Sensing of Environment* 150: 172–187. doi:10.1016/j.rse.2014.05.001.

Roelfsema, C. M., and S. R. Phinn. 2009. *A Manual for Conducting Georeferenced Photo Transects Surveys to Assess the Benthos of Coral Reef and Seagrass Habitats*. Queensland: Centre for Remote Sensing & Spatial Information Science, School of Geography, Planning & Environmental Management, University of Queensland.

Van der Meer, F. D., and S. de Jong, Eds. 2001. *Imaging Spectrometry; Basic Principles and Prospective Applications*. Dordrecht, The Netherlands: Kluwer Academic Publishers.

Vapnik, V. N. 1995. *The Nature of Statistical Learning Theory*. New York: Springer-Verlag.

Wahidin, N., V. P. Siregar, B. Nababan, I. Jaya, and S. Wouthuyzen. 2015. "Object Based Image Analysis for Coral Reef Benthic Habitat Mapping with Several Classification Algorithm." *Procedia Environmental Sciences* 24: 222–227. Elsevier. doi:10.1016/j.proenv.2015.03.029.

Wicaksono, P. 2010. *Integrated Model of Water Column Correction Technique for Improving Satellite-Based Benthic Habitat Mapping, A Case Study on Part of Karimunjawa Islands, Indonesia*. Yogyakarta: Faculty of Geography, Universitas Gadjah Mada.

Wicaksono, P. 2015. *Remote Sensing Model Development for Seagrass and Mangrove Carbon Stock Mapping*. Yogyakarta: Faculty of Geography, Universitas Gadjah Mada.

Wicaksono, P. 2016. "Improving the Accuracy of Multispectral-Based Benthic Habitats Mapping Using Image Rotations: The Application of Principle Component Analysis and Independent Component Analysis." *European Journal of Remote Sensing* 49: 433–463. doi:10.5721/EuJRS20164924.

Wicaksono, P., I. S. Kumara, M. Kamal, M. A. Fauzan, Z. Zhafarina, D. A. Nurswantoro, and R. N. Yogyantoro. 2017. "Multispectral Resampling of Seagrass Species Spectra: WorldView-2,

Quickbird, Sentinel-2A, ASTER VNIR, and Landsat 8 OLI." *IOP Conference Series: Earth and Environmental Science* 98 (2017): 012039. Bristol, UK: IOP Publishing. doi:10.1088/1755-1315/98/1/012039.

Wicaksono, P., and M. Hafizt. 2013. "Mapping Seagrass from Space: Addressing the Complexity of Seagrass LAI Mapping." *European Journal of Remote Sensing* 46: 18–39. doi: 10.5721/EuJRS20134602.

Wicaksono, P., and M. Hafizt. 2018. "Dark Target Effectiveness for Dark-Object Subtraction Atmospheric Correction Method on Mangrove Above-Ground Carbon Stock Mapping." *IET Image Processing* 12 (4): 582–587. doi:10.1049/iet-ipr.2017.0295.

Zhang, C. 2015. "Applying Data Fusion Techniques for Benthic Habitat Mapping and Monitoring in a Coral Reef Ecosystem." *ISPRS Journal of Photogrammetry and Remote Sensing* 104: 213–223. doi:10.1016/j.isprsjprs.2014.06.005.

Zhang, C., D. Selch, Z. Xie, C. Roberts, H. Cooper, and G. Chen. 2013. "Object-Based Benthic Habitat Mapping in the Florida Keys from Hyperspectral Imagery." *Estuarine, Coastal and Shelf Science* 134: 88–97. doi:10.1016/j.ecss.2013.09.018.

Index

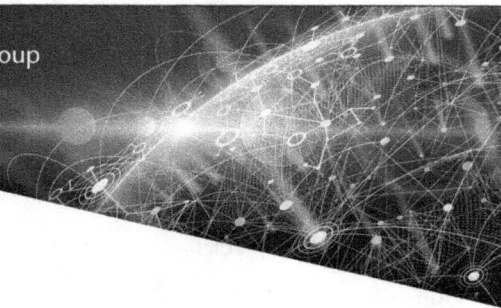

For Product Safety Concerns and Information please contact our EU
representative GPSR@taylorandfrancis.com
Taylor & Francis Verlag GmbH, Kaufingerstraße 24, 80331 München, Germany

www.ingramcontent.com/pod-product-compliance
Lightning Source LLC
Chambersburg PA
CBHW081106220326
41598CB00038B/7249